Intelligent Approaches to Cyber Security

Intelligent Approaches to Cyber Security provides details of important cyber security threats and their mitigation and the influence of machine learning (ML), deep learning (DL) and blockchain technologies in the realm of cyber security.

Features:

- Role of deep learning and machine learning in the field of cyber security
- Using ML to defend against cyber-attacks
- Using DL to defend against cyber-attacks
- Using blockchain to defend against cyber-attacks

This reference text will be useful for students and researchers interested and working in the area of future cyber security issues in the light of emerging technology in the cyber world.

Intelligent Approaches to Cyber Security

Edited by
Narendra M. Shekokar
Hari Vasudevan
Surya S. Durbha
Antonis Michalas
Tatwadarshi P. Nagarhalli

CRC Press
Taylor & Francis Group
Boca Raton London New York

CRC Press is an imprint of the
Taylor & Francis Group, an **informa** business

A CHAPMAN & HALL BOOK

Designed cover image: @Shutterstock

First edition published 2024
by CRC Press
6000 Broken Sound Parkway NW, Suite 300, Boca Raton, FL 33487-2742

and by CRC Press
4 Park Square, Milton Park, Abingdon, Oxon, OX14 4RN

CRC Press is an imprint of Taylor & Francis Group, LLC

ISBN: 978-1-032-52161-9 (hbk)
ISBN: 978-1-032-52769-7 (pbk)
ISBN: 978-1-003-40830-7 (ebk)

DOI: 10.1201/9781003408307

Typeset in Palatino
by SPi Technologies India Pvt Ltd (Straive)

Contents

Section I   Introduction to Machine Learning in Cyber Security

Section II   Defending Against Cyber Attack Using Machine Learning

Section III   Defending Against Cyber Attack Using Deep Learning

Section IV   Defending Against Cyber Attack Using Advance Technology

Preface

The number of internet users has increased dramatically during the last few decades. The improvement in internet speed and the drop in the price of using it has contributed to this expansion. With the advent of the internet, billions of individuals are now able to go online. The growth of the internet has given rise to a number of new industries, like social media and online financial services, which have been effective in attracting online customers. In addition, new technologies like the Internet of Things and the Industrial Internet of Things have emerged as a result of the high-speed internet boom.

The whole ecosystem of government services and governance may now be found online, thanks to the high-speed internet. The internet is now widely used in both the military and the healthcare industries. Wearable healthcare devices have been developed as a result of combining the Internet of Things with the healthcare industry and these are useful for monitoring the health of those who use them.

As the internet is a very open and unprotected method of communication, it has attracted numerous miscreants with nefarious purposes. These criminals have made repeated attempts to get information and misuse it. Hence, the number of cyber-attacks and cybercrimes has increased exponentially with the growth of high-speed internet users. Cyber security is a significant issue in this setting that needs in-depth research and analysis.

The advent of technologies like machine learning, deep learning, artificial intelligence and blockchain have brought about a paradigm shift in the way security issues are handled. Incorporating these technologies in different aspects of cyber security has increased the robustness of various security mechanisms. Hence, the main aim of the book is to study the impact of these technologies on cyber security.

Authentication and access control are ways in which a layer of security can be implemented, and, hence, studying the same is important. The employment and the impact of state-of-the-art technologies like machine learning, deep learning, artificial intelligence and blockchain in different fields are also studied and covered.

The book is divided into a number of sections to arrange its contents effectively. The themes discussed vary from the requirements of technologies like machine learning, deep learning and blockchain in many areas to the usefulness of these technologies in cyber security. The book also offers new developments and research concerns relevant to the subject.

Hence, editors have worked to achieve completeness and coverage of the domain's range.

Eleven chapters make up the book, and address themes like comprehending the importance of machine learning and deep learning, as well as how these technologies – along with blockchain – can be critical to many aspects of cyber security.

Editors

Dr Narendra M. Shekokar earned his PhD in Engineering (Network Security) from NMIMS University, Mumbai and he is a professor in the Department of Computer Engineering and Head of the Department of IoT and Cyber Security with Blockchain Technology at SVKM's Dwarkadas J. Sanghvi College of Engineering, Mumbai (an autonomous college affiliated to the University of Mumbai). He was a member of the Board of Studies at the University of Mumbai for more than five years, is currently on department advisory boards at various institutes, and has also been a member of various commit-tees at the University of Mumbai. He has 25 years of teaching experience.

Dr Shekokar has guided six research fellows and currently has six research fellows registered with him. He has also guided 26 students at the post-graduate level. He has presented more than 65 papers at international and national conferences and has also published more than 20 research papers in renowned journals. He is editor of two renowned books in the ML, DL and cyber security domains published by Taylor & Francis, CRC Press, USA. He has received the minor research grant twice from the University of Mumbai for his research projects. He has delivered expert talks and chaired sessions at numerous events and international conferences.

Dr Hari Vasudevan is the principal of SVKM's Dwarkadas J. Sanghvi College of Engineering, Mumbai, an autonomous institution affiliated to the University of Mumbai. He has been the prin-cipal of the college since February 2009. He has a PhD degree from IIT Bombay and a Master of Engineering degree, as well as a postgradu-ate Diploma in Industrial Engineering from VJTI, University of Mumbai.

Dr Vasudevan is a certified ERP consultant from the S.P. Jain Institute of Management and Research, Mumbai, under the University Synergy Program of the BaaN Institute, the Netherlands. He is a member of the Research and Recognition Committee (RRC) of the University of Mumbai in mechanical engineering as the internal expert of the university. He is also the chairman of the board of studies in production engineering at the University of Mumbai. He has over 29 years of experience in teaching and two years of experience in industry and his areas of interest include: digital manufacturing, cyber security in manufacturing and manufacturing strategy. He has published over 132 research papers in international journals and conferences as well as in national journals and conferences and has published textbooks, textbook chapters, workbooks and articles in various newsletters. He is a recognized PhD guide at the University of Mumbai as well as at the NMIMS (deemed to be a university) in the field of mechanical engineering. Eight students have so far earned their PhD degrees under his guidance. He is currently guiding five more PhD students at the DJSCE research center of the University of Mumbai. He has six Indian design and two process patents to his credit and has received many awards and accolades from various national level organizations in the past. Dr Hari Vasudevan is a fellow of the Institution of Engineers, India and is currently the president and fellow of the Indian Society of Manufacturing Engineers, Mumbai.

Dr Surya S. Durbha earned his PhD in Computer Engineering from Mississippi State University (MSU), USA. He is a professor at the Centre of Studies in Resources Engineering (CSRE), Indian Institute of Technology Bombay (IIT-Bombay). He has more than 20 years of experience.

He has won a number of awards, including the best paper award in an international conference on geoinfomatics, Outstanding Research award (HM) (2008), GRI, Mississippi State University, State Pride faculty award (2010), Mississippi State University, Excellence in Teaching award, IIT Bombay and NVIDIA Innovation award by NVIDIA.

He is a reviewer for many prestigious journals including IEEE Geoscience and Remote Sensing, IEEE Journal of Selected Topics in Applied Earth Observations and Remote Sensing, and Geoinformatica, among others. He has guided more than six PhD and twenty-four MTech students. He has also had many research papers published in prestigious journals and conferences.

Dr Antonis Michalas earned his PhD in Network Security from Aalborg university, Denmark and currently he is an assistant professor in the Department of Computing Science at Tampere University of Technology, in the Faculty of Computing and Electrical Engineering. Prior to this, he was an assistant professor in cyber security at the University of Westminster, London. Earlier, he was a postdoctoral researcher at the security lab at the Swedish Institute of Computer Science in Stockholm, Sweden. As a postdoctoral researcher at the SCE Labs, he was actively involved in national and European research projects.

Dr Michalas has published a significant number of papers in field-related journals and conferences and has also participated as a speaker in various conferences and workshops. His research interest includes private and secure e-voting systems, reputation systems, privacy in decentralized environments, cloud computing, trusted computing and privacy preserving protocols in participatory sensing applications.

Dr Tatwadarshi P. Nagarhalli is an associate professor and head of the Department of Artificial Intelligence and Data Science at Vidyavardhini's College of Engineering and Technology, Vasai, Mumbai. He has two PhD degrees, in Computer Engineering and Sanskrit. He has over ten years of experience, including two years of industry experience. His areas of interest include: data security, machine learning, natural language processing, and artificial intelligence. He has taken subjects like machine learning, advance system security and digital forensics, artificial intelligence, natural language processing, applied data science and mathematics in AI-ML. He has served as a panel member for the syllabus design of machine learning, deep learning, natural language processing and artificial intelligence at the University of Mumbai.

He has presented and published more than 55 research papers in reputed conferences and journals (SCI and Scopus Indexed) like IEEE and Springer. He has served as an editor for the book titled *Cyber Security Threats and Challenges Facing Human Life* published by CRC Press, Taylor & Francis group. He has also authored many book chapters with reputed publishers. He has received the Best Paper Award for a data security paper at an international IEEE conference. He has had two patents published, one of which is on data security. He has five software/literary copyrights to his credit as well. He is the founder and has worked as the editor-in-chief of a non-profit international journal indexed in a few reputed databases. He is also an active reviewer of reputed Scopus indexed journals like the IEEE Access and Journal of King Saud University – Computer and Information Science. He is a lifetime member of the Indian Society for Technical Education. He has undertaken many workshops and training programs at undergraduate and postgraduate levels for industry and academia.

Contributors

Smita Bansod
Shah and Anchor Kutchhi
  Engineering College
University of Mumbai
Mumbai, India

Harshita Bhagwat
Alamuri Ratnamala Institute of
  Engineering and Technology
Shahapur, India

Priyanka Bhatele
School of Computer Engineering
  and Technology
MIT World Peace University
Pune, India

Gresha Bhatia
Vivekanand Education Society's
  Institute of Technology
University of Mumbai
Mumbai, India

Bhavi Dave
Dwarkadas J. Sanghvi College of
  Engineering
Mumbai, India

Deep Gandhi
Dwarkadas J. Sanghvi College of
  Engineering
University of Mumbai
Mumbai, India

Aruna Gawade
Dwarkadas J. Sanghvi College of
  Engineering
Mumbai, India

Uttara Gogate
Shivajirao S. Jondhale College of
  Engineering
University of Mumbai
Mumbai, India

Amruta Hingmire
MIT World Peace University
Pune, India

Swati Jadhav
MIT World Peace University
Pune, India

Jyoti Khurpade
MIT World Peace University
Pune, India

Vanita Mane
Ramrao Adik Institute of Technology,
Mumbai, India

Monika Mangla
Dwarkadas J. Sanghvi College of
  Engineering
University of Mumbai
Mumbai, India

Ramchandra Mangrulkar
Dwarkadas J. Sanghvi College of
  Engineering
University of Mumbai
Mumbai, India

Jyoti Mante
School of Computer Engineering
  and Technology
MIT World Peace University
Pune, India

Kunal Mohan Meher
Xavier Institute of Engineering
Mumbai, India

Jash Mehta
Dwarkadas J. Sanghvi College of
    Engineering
University of Mumbai
Mumbai, India

Tatwadarshi P. Nagarhalli
Vidyavardhini's College of
    Engineering and Technology
University of Mumbai, India

Snehal Paddalwar
Ramrao Adik Institute of Technology
Mumbai, India

Rohini Patil
Terna Engineering College
University of Mumbai, India

Sharvari Patil
Dwarkadas J. Sanghvi College of
    Engineering
University of Mumbai, India

Uma Pujeri
MIT World Peace University
Pune, India

Leena Ragha
Ramrao Adik Institute of
    Technology
Mumbai, India

Ashwini M. Save
Dwarkadas J. Sanghvi College of
    Engineering
University of Mumbai
Mumbai, India

Narendra M. Shekokar
Dwarkadas J. Sanghvi College of
    Engineering
University of Mumbai
Mumbai, India

Pallavi Shimpi
MIT World Peace University
Pune, India

Swati Sinha
MIT World Peace University
Pune, India

Aditi Vora
Dwarkadas J. Sanghvi College of
    Engineering
Mumbai, India

Section I

Introduction to Machine Learning in Cyber Security

1

Introduction and Importance of Machine Learning Techniques in Cyber Security

Gresha Bhatia

Vivekanand Education Society's Institute of Technology, University of Mumbai, Mumbai, India

Narendra M. Shekokar

Dwarkadas J. Sanghvi College of Engineering, University of Mumbai, Mumbai, India

Tatwadarshi P. Nagarhalli

Vidyavardhini's College of Engineering and Technology, University of Mumbai, Mumbai, India

CONTENTS

1.1 Introduction

Machine learning (ML) is a buzzword in today's world. Every application is being built, rebuilt and reinvented keeping machine learning in mind. This thought process gained impetus when programs were written in a manner different from the traditional approach and according to the demands and requirements of the users.

The traditional approach consisted of developing systems through a set of instructions. The instructions usually consisted of an IF-THEN-ELSE structure, where when certain conditions were met, the program would execute a

DOI: 10.1201/9781003408307-2

specific action. These instructions gathered different sets of input parameters, enabled the computer to process them and further transform them into the desired output. The outputs obtained would either be stored in the memory for further use or displayed to the user.

Machine learning, on the other hand, is said to be an advancement in the way computer programs are written. Machine learning programs are automated processes that enable machines to self-learn, solve problems with little or no human input and take actions based on past observations. Therefore, ML programs learn by themselves, without being explicitly programmed from the existing (training) data and improve themselves over time [1].

For example, once machines are taught to differentiate between apples and pears, by showing them examples of fruit, the machine will automatically start identifying and labeling fruits by themselves. This would only be possible if they have learned to differentiate between the fruits through proper datasets and accurate training examples [2].

Thus, machine learning programs can be put to work for a number of applications [3] such as automated translation [4], image recognition, traffic prediction, product recommendations [5], chatbot development [6], email spam and malware filtering, voice search technology, emotion detection [7], in the medical domain [8–10], self-driving cars, online fraud detection, identifying cyber bullies [11, 12], prediction and identification of diseases in plants [13], cyber security and many more. These applications utilize various techniques in terms of the supervised, unsupervised, reinforcement and semi-supervised techniques of ML [14].

1.2 Importance of ML Techniques in Cyber Security

Cyber security is considered to be a set of technologies, processes and controls that are applied to protect systems, networks, programs, devices and data from cyber attacks. Cyber attacks are said to be automated attacks that aim to exploit and destroy the smooth functioning of the systems.

1.2.1 Common Vulnerabilities

Common cyber threats include [15, 16]:

- Malware, such as ransomware, botnet software, remote access Trojans, rootkits and bootkits, spyware, Trojans, viruses and worms.
- Backdoors, which allow remote access.
- Formjacking, which inserts malicious code into online forms.
- Cryptojacking, which installs illicit cryptocurrency mining software.

- DDoS (distributed denial-of-service) attacks which flood servers, systems and networks with traffic.
- DNS (domain name system) poisoning attacks which compromise the DNS to redirect traffic to malicious sites.
- Phishing, which includes baiting the victims into revealing personal information.

Since there are cyber-attacks, there are mechanisms to protect the systems. One way of dealing with them is the utilization of machine learning and deep learning to protect the network as well as the systems. These cyber-attacks can be prevented by blocking malware and phishing threats as well as identifying vulnerabilities and the utilization of machine learning and deep learning forms one of the main pillars that guarantees data and information security.

It is also important to note that these machine learning and deep learning techniques play a very important role in identifying and mitigating cyber-attacks. Some of the attacks and their machine learning/deep learning counters are discussed in this book.

1.2.2 Assets That Need to Be Protected

Some of the assets that require an active protection mechanism include:

1. **Critical infrastructure**: The critical infrastructure in an organization is considered to be most vulnerable to attack as compared to other systems that rely on older software. These organizations need to implement appropriate measures to manage the security risks.
2. **Network**: Networks within the organization, such as the operating systems, servers, hosts, firewalls, wireless access points, network protocols and so on, need to be made secure from cyber-attacks [17].
3. **Cloud**: Securing data, applications and infrastructure in the cloud is also an important way for assets to be protected [18].
4. **IoT (Internet of Things)**: Devices connected to the internet also need to be secured. If such devices are connected in an IoT environment, this would require securing these smart devices and networks.
5. **Applications**: Vulnerabilities resulting from insecure development processes in designing and coding must also be taken care of.

1.2.3 Role of Machine Learning in Cyber Security

In order to provide security, the cyber security system needs to identify and analyze the symptoms. These can be used further to learn similar types of patterns observed previously. They can then be utilized to prevent similar attacks and respond to changing behavior. Machine learning algorithms can be trained to perform these tasks, learn and optimize pattern identification, respond to active attacks in real time, develop patterns and manipulate those patterns with algorithms. To perform all of this, the underlying data has to be complete, relevant, rich and accurate to represent as many potential outcomes from as many potential scenarios as possible. The data has to be rich enough to provide details about machines, applications, protocols and network sensors. This rich data can then be used to build different models and different aspects of the behavior. Various algorithms can then be developed to make decisions about when to issue alerts, when to take action to respond to potential threats and so forth.

The primary objective of developing and applying machine learning in cyber security is to make the process of malware detection more actionable, scalable and effective than traditional approaches. Support vector machines and Bayesian classification techniques of machine learning help in mitigating cyber-attacks, detect hidden trends and build a data-driven machine learning model to prevent attacks. Similar to this, many machine learning and deep learning algorithms can be effectively used in different arenas of cyber security.

Machine learning techniques used for cybersecurity include:

- **Supervised learning**: In this, the model relies on the patterns available in the dataset. It then uses these internal patterns to group the data into different categories. Examples include: K-means, sequential pattern mining, DB scan, a priori algorithm, classification and regression methods. These task-based methods are used for classifying or predicting the target variable for a particular security threat, denial-of-service (DoS) attack (yes, no) or identifying distinct labels of network risks, such as scanning and spoofing.

- **Unsupervised learning**: The aim of unsupervised learning, or data-driven learning is to uncover patterns, structures or relevant information in unlabeled data. Clustering techniques can be utilized to discover hidden patterns, spot anomalies and policy violations.

1.3 Stages of a Cyber-Attack

Cyber-attacks are divided into five phases. These include [19]:

1. **Reconnaissance or preparation phase of the attack**: In this the attacker utilizes phishing or malicious calls, also known as social engineering attacks. Machine learning algorithms can look for email signatures

and voice phishing attacks, detect malicious or phishing email signatures and flag and block them. Machine learning can further be used to scan any external devices that are connected and prevent malicious software from being moved to the computers of the organization. Rule-based machine learning algorithms can be used to recommend a list of passwords that can be used to prevent unauthorized access.

2. **Scan or weaponization**: In this phase, the attacker exploits the vulnerabilities of the target system using automated tools. Learning-based penetration tests using ML algorithms can be used to automatically determine the weaknesses prior to the attacks.

3. **Attack**: Attacks such as spam detection, malware detection, denial-of-service attacks, network anomaly detection, identity theft detection, information leakage detection and social media analytics can be prevented through use of ML algorithms. A number of algorithms such as linear regression, polynomial regression, logistic regression, naïve Bayes classifier, support vector machine, decision tree, nearest neighbor, clustering, dimensionality reduction, linear discriminant analysis and boosting can be used to provide cyber security.

4. **Maintain access**: In this phase, malware such as Trojans, backdoor or emotions are used by the attacker to maintain access to the systems. Machine learning algorithms such as support vector machines can be used to detect malware traffic packets when the malware contacts the attacker and vice versa. Various clustering algorithms such as K-means, DBSCAN and Hierarchical can also be used for malware, phishing attacks, side-channel attack detection and spam filtering.

5. **Cover tracks and hiding**: In this phase, the attacker specifies a requirement not to be tracked. Deceptive training data make the algorithm inefficient. This process of forging training data is called adversarial machine learning (AML). Improved machine learning techniques can act as a measure against adversarial attacks.

Similar to how these learning techniques can be used for aiding at different stages of cyber-attacks, these machine learning and deep learning techniques can also be used for effective mitigation at every stage of the cyber-attack.

1.4 Conclusion

Securing cyber space is an important job that needs special attention. With the advent of intelligent technologies like machine learning and blockchain, the attackers have been become smarter. Nonetheless, these technologies can be effectively leveraged into making defense mechanisms smarter as well.

This chapter focused on how machine learning and deep learning techniques can be effectively used at different levels in the cyber security space, from different threats and attacks to stages in cyber security attacks, these learning approaches have a very important role to play. Apart from these learning approaches and techniques, other recent technologies like blockchain also play an important role in the field of cyber security.

References

[1] E. Alpaydın, *Introduction to Machine Learning*, 2nd Edition, The MIT Press, 2010.
[2] Aishwarya Sarkale, Kaiwant Shah, Anandji Chaudhary, and Tatwadarshi P. Nagarhalli, "An Innovative Machine Learning Approach for Object Detection and Recognition", *IEEE Second International Conference on Inventive Communication and Computational Technologies (ICICCT)*, 2018, pp. 1008–1010.
[3] Ramchandra Sharad Mangrulkar, Antonis Michalas, Narendra Shekokar, Meera Narvekar, and Pallavi Vijay Chavan, *Design of Intelligent Applications Using Machine Learning and Deep Learning Techniques*, edited by Ramchandra Sharad Mangrulkar, Antonis Michalas, Narendra Shekokar, Meera Narvekar, and Pallavi Vijay Chavan, Chapman and Hall/CRC, 2021.
[4] Tatwadarshi P. Nagarhalli, Sneha Mhatre, Sanket Patil, and Prafulla Patil, "The Review of Natural Language Processing Applications with Emphasis on Machine Learning Implementations", *IEEE International Conference on Electronics and Renewable Systems (ICEARS)*, 2022, pp. 1353–1358.
[5] Salil Kanetkar, Akshay Nayak, Sridhar Swamy, and Gresha Bhatia, "Web-Based Personalized Hybrid Book Recommendation System", *International Conference on Advances in Engineering & Technology Research (ICAETR-2014)*, 2014, pp. 1–5.
[6] Tatwadarshi P. Nagarhalli, Vinod Vaze, and N. K. Rana, "A Review of Current Trends in the Development of Chatbot Systems", *IEEE 6th International Conference on Advanced Computing & Communication Systems (ICACCS)*, 2020, pp. 706–710.
[7] Ved Kokane, Prasad Nijai, Vikas Jamge, and Tatwadarshi P. Nagarhalli, "Speech Emotion Recognition Using Convolutional Neural Networks and Long Short-Term Memory", *IEEE 6th International Conference on Trends in Electronics and Informatics (ICOEI)*, 2022, pp. 1–8.
[8] Tatwadarshi P. Nagarhalli, "Analysis of Recurrent Neural Network and Convolution Neural Network Techniques in Blood Cell Classification", *Next Generation Healthcare Systems Using Soft Computing Technique*, edited by Rekh Ram Janghel, Rohit Raja, Korhan Cengiz, and Hiral Raja, Chapman and Hall/CRC, 2022.
[9] Tatwadarshi P. Nagarhalli, Sneha Mhatre, Ashwini Save, and Sanket Patil, "Evaluating the Effectiveness of the Convolution Neural Network in Detecting Brain Tumors", *Next Generation Healthcare Systems Using Soft Computing Technique*, edited by Rekh Ram Janghel, Rohit Raja, Korhan Cengiz, and Hiral Raja, Chapman and Hall/CRC, 2022.

[10] Smita Sanjay Ambarkar and Narendra Shekokar, "Toward Smart and Secure IoT Based Healthcare System", *Internet of Things, Smart Computing and Technology: A Roadmap Ahead*, Springer International Publishing, 2020, pp. 283–303.

[11] Akankshi Mody, Shreni Shah, Reeya Pimple, and Narendra Shekokar, "Identification of Potential Cyber Bullying Tweets Using Hybrid Approach in Sentiment Analysis", *International Conference on Electrical, Electronics, Communication, Computer, and Optimization Techniques (ICEECCOT)*, 2014, pp. 878–881.

[12] Krishna B. Kansara and Narendra M. Shekokar, "A Framework for Cyberbullying Detection in Social Network", *International Journal of Current Engineering and Technology*, Vol. 1, Issue 5, 2015, pp. 494–498.

[13] Gresha S. Bhatia, Pankaj Ahuja, Devendra Chaudhari, Sanket Paratkar, and Akshaya Patil, "Plant Disease Detection Using Deep Learning", *Second International Conference on Computer Networks and Communication Technologies: ICCNCT 2019*, 2020, pp. 408–415.

[14] Tatwadarshi P. Nagarhalli, Ashwini Save, and Narendra Shekokar, "Fundamental Models in Machine Learning and Deep Learning", *Design of Intelligent Applications using Machine Learning and Deep Learning Techniques*, edited By Ramchandra Sharad Mangrulkar, Antonis Michalas, Narendra Shekokar, Meera Narvekar and Pallavi Vijay Chavan, Chapman and Hall/CRC, 2021.

[15] N. M. Shekokar, C. Shah, M. Mahajan, and S. Rachh, "An Ideal Approach for Detection and Prevention of Phishing Attacks", *Procedia Computer Science*, Vol. 49, 2015, pp. 82–91.

[16] Sharvari Prakash Chorghe and Narendra Shekokar, "A Survey on Anti-Phishing Techniques in Mobile Phones", *International Conference on Inventive Computation Technologies (ICICT)*, 2016, pp. 1–5.

[17] M. D. Shah, S. N. Gala, and N. M. Shekokar, "Lightweight Authentication Protocol Used in Wireless Sensor Network", *2014 International Conference on Circuits, Systems, Communication and Information Technology Applications (CSCITA)*, IEEE, 2014, April, pp. 138–143.

[18] N. Shekokar, K. Sampat, C. Chandawalla, and J. Shah, "Implementation of Fuzzy Keyword Search over Encrypted Data in Cloud Computing", *Procedia Computer Science*, Vol. 45, 2015, pp. 499–505.

[19] Shivam Arora, "Explore the 5 Phases of Ethical Hacking". https://www.simplilearn.com/phases-of-ethical-hacking-article (Accessed on 21st January, 2023).

2

Review of Machine Learning Approaches in the Field of Healthcare

Uttara Gogate
University of Mumbai, Mumbai, India

Harshita Bhagwat
Alamuri Ratnamala Institute of Engineering and Technology, Shahapur, India

CONTENTS

2.1 Introduction

Machine learning (ML) comes within the scope of artificial intelligence, in which machines replicate people's abilities. Machine learning is very well known in data processing [1] and ML techniques are based on two conditions: (1) testing data, and (2) training data. The system receives training openly from records and experiences. Using training data, tests are applied on dissimilar types of data and then requisite ML algorithms are applied.

The primary aim is to allow the computers to learn automatically, either from available types of data or from past experiences, without human intervention or assistance and adjust its actions accordingly.

There are several application areas of machine learning. Many researchers have successfully developed and implemented different working models of the following ML applications:

- Online fraud detection
- Traffic prediction
- Image recognition
- Speech recognition
- Self-driving cars
- Email spam and malware filtering
- Virtual personal assistant
- Automatic language translation

Though there are several application areas of machine learning, recently, its use in healthcare is developing far and wide and is helping patients and clinicians in many different ways. The most common uses of machine learning in the healthcare domain are in automating medical billing, clinical decision support and the development of clinical care guidelines [2]. More attention is being drawn to the automatic decision support systems which help physicians to diagnose any health abnormality accurately and to detect any disease in earlier stages. We will review some of the common diseases and abnormalities, where machine learning techniques are found to be very useful in accurate and appropriate diagnostic and decision making.

The modern healthcare system also faces major security threats due to highly sensitive data usage. There are numerous cases where we need to attend to different security issues like network issues, human intruders, data theft and many more. We will focus on the most common security issues of ML in the healthcare domain.

2.2 Machine Learning Algorithms

Machine learning algorithms are broadly categorized in following four types:

1. Supervised machine learning algorithms
2. Unsupervised machine learning algorithms

3. Semi-supervised machine learning algorithms

4. Reinforcement machine learning algorithms

These algorithms are further classified into different methods as shown in Figure 2.1 and used according to the requirements of the application.

In the healthcare domain, different ML algorithms are used to solve the five typical teething troubles which healthcare givers/practitioners have to manage. These can be responded to with the help of different machine learning algorithms as listed in Table 2.1.

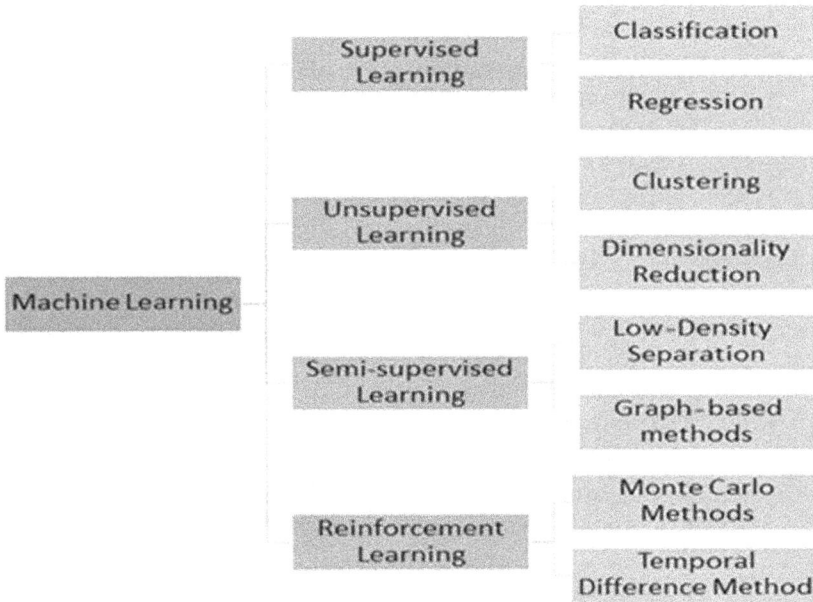

FIGURE 2.1
Classification of ML algorithms [3].

TABLE 2.1

Machine Learning in the Healthcare Domain

Question/Problem in Healthcare	ML Algorithm Used to Solve
Is this disease A or B or other?	Classification Algorithm
Are the vital signs unusual?	Anomaly Detection Algorithm
How much or how many? (count of cells/vitals)	Regression Algorithm
How are the patients' records organized?	Clustering Algorithm
What action/treatment should be the next?	Reinforcement Learning

2.3 Machine Learning Models

Machine learning involves creating appropriate models, based on the different problems requiring solutions, which are trained on some training data and can then process additional data to make predictions. Various models based on different types of algorithms such as supervised, unsupervised or reinforcement learning as shown in Figure 2.1 have been used and explored for machine learning and are useful in healthcare systems [4].

I. **Supervised Learning Algorithms**

Some important supervised learning algorithms which are commonly used in ML based healthcare systems are discussed first:

Artificial Neural Network (ANN): As shown in Figure 2.2(a), ANN is a parallel system which provides absolute, discrete and vector value function which is familiar and a realistic approach in the training phase. It is useful for learning valid prized, distinct prized and vector valued functions. This network is similar in nature to a human neural system which has neuron elements working to solve definite problems [5]. So, ANN gives very good results in solving classification problems in healthcare applications but is hardly ever used for predictive analysis.

Decision Tree Model (DT): The structure of a DT is a tree-shaped graph structure and is used as a classifier. It consists of three nodes: (1)root nodes, (2) leaf nodes, and (3) internal nodes. The internal node indicates attribute test, the leaf node indicates class distribution and the root/origin node is the main node of the tree [5]. The DT includes concepts such as node connection and labels as shown in Figure 2.2(b) [6]. The DT algorithm helps in making correct decisions based on some medical conditions given and, hence, is useful for solving regression as well as classification problems.

Support Vector Machine (SVM): An SVM is a supervised ML technique that helps in the distribution of a variety of data by sorting the hyper plane. The SVM creates the hyper plane in an elevated dimensional gap. In this, the samples of training data are divided into positive and negative hyper planes [5]. The SVM can be used effectively for solving classification as well as regression type of problems in different healthcare applications.

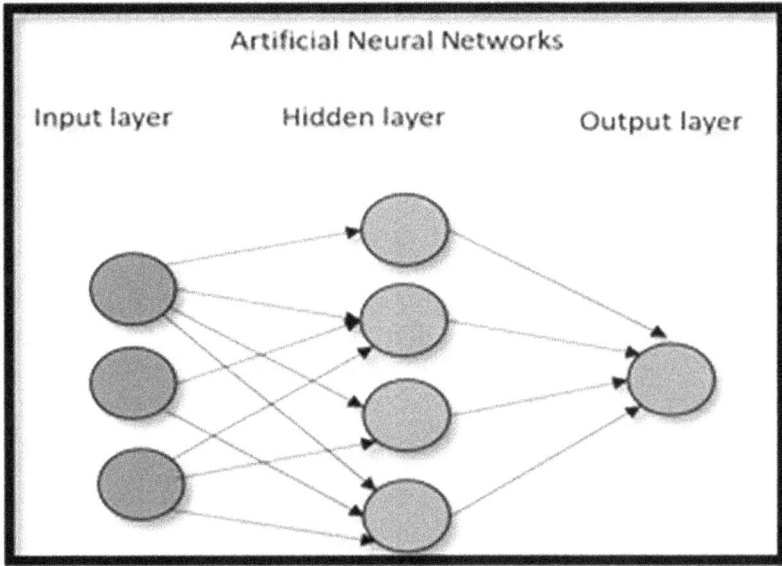

(a)

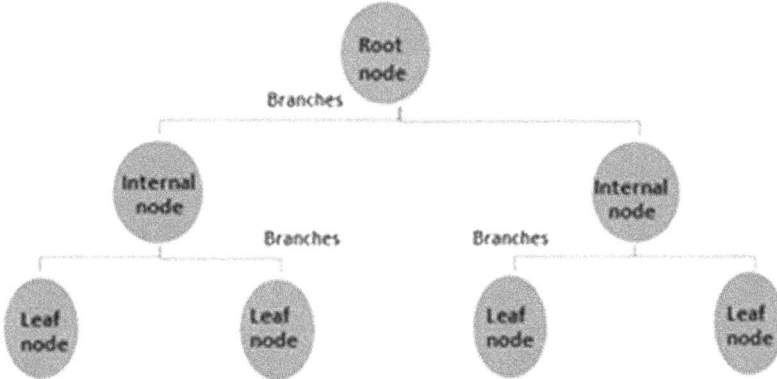

(b)

FIGURE 2.2
(a) ANN model and (b) Decision Tree model.

Linear Regression (LR): Linear regression usually tries to find the best-fit line, which can be used to predict the output more accurately. Therefore, supervised machine learning algorithms consisting of the regression concept are effectively used in disease prediction. The LR concepts are easy to understand and useful for over-fitting by regularization. The limitation of the LR concept is that it is not a good fit for nonlinear relationships [6]. In regression we require that the output variable should always be continuous or real valued.

Logistic Regression: Logistic regression is popularly used in classification problem. It provides the probability results (between 0–1) based on input values and, hence, works best on binary classification problems. In healthcare, logistic regression is mainly used to solve YES/NO types of problems such as "infected or not infected?" [7].

Naïve Bayes: Naïve Bayes is a family of probabilistic ML algorithms based on Bayes theorem and can be used in solving wide-ranging classification problems. The Naïve Bayes method performs well with huge records which consist of binary and multi-class data. It is mainly used for analyzing text and processing natural language data [8]. The main advantages of using the Naïve Bayes algorithm is that it doesn't need a large quantity of training data. Also, it can handle both continuous as well as discrete data. It is found to be highly scalable with the number of predictors and data points. It gives fast decisions and, hence, can be used to make predictions in real-time.

K-Nearest Neighbor (K-NN): This can find the distance between the nearest data labels of the training data and a test data set using K values. It is simple and flexible with attributes and distance functions. It can support multi-class data sets and can be used for solving both regression and classification problems. For classification problems, K-NN is the best choice as it is a supervised learning algorithm [8]. Although K-NN is the best choice in many applications, including healthcare, due to its simplicity, its major drawback is that the execution speed becomes remarkably slow with growing data sizes.

Supervised learning is popularly used for solving classification and regression-based healthcare issues. Nonetheless, unsupervised and reinforcement learning algorithms are also extensively used in the healthcare domain, as discussed in the following sections.

II. Unsupervised Learning Algorithms

Hierarchical Clustering: Clustering is a type of unsupervised machine learning algorithm used to create clusters based on different features, which subsequently improve the classification model and its accuracy. Different clustering techniques have been widely used in the healthcare domain for easy and faster diagnosis and prediction of various diseases. It provides quick, satisfactory, trustworthy and cost-effective healthcare provisions to the patients. In hierarchical clustering each item has one cluster. Thus, for N data items there are N clusters. So, the distance (similarities calculated) between the item and the cluster is similar [9].

Partitioning Clustering: This is also an unsupervised type of learning algorithm. Unlike in supervised learning, there is no labeled data used for this type of clustering. It consists of two types – K-Means and Fuzzy C-means:

- *K-Means*: This considers the similarities for the division of objects to form clusters. This means the features belonging to two different clusters cannot be similar. They possess dissimilar features. Here, K indicates the number of clusters to be created.
- *Fuzzy C-means (FCM)*: Unlike K-means, this method of clustering allows one object of data to belong to more than one cluster.

The clustering methods discussed here possess an automatic recovery mechanism from any kind of failure, without any human intervention. This enhances the availability and reliability of the system which is one of the most essential features of any healthcare system. But, at the same time it adds complexity to the system. The major disadvantage is the inability of the system to be recovered from database corruption [10].

III. **Reinforcement Learning**

The next category of ML algorithm is reinforcement learning where a machine learns from the rewards of feedback and also improves its future results. The policy of reinforcement learning from its agents can be of following two types:

- *On Policy*: the agent learns from the current action that is delivered from the currently used policy, e.g., a SARSA algorithm which learns Q-values from the actions performed by current policy.
- *Off Policy*: the agent learns from the action that is delivered from any other policy used, e.g., a Q-learning algorithm which learns Q-values using a 'greedy' approach [11].

Q-Learning: Q-learning is a value-based reinforcement learning model that uses a bellman equation for prediction of data. It takes action after receiving data from another policy, so it is called an off-policy algorithm [12]. A Q-table is used to find the best action for each learning.

SARSA (State Action Reward State Action): SARSA and Q-learning are very similar concepts. Instead of using a greedy policy like Q-learning, SARSA learns the Q-value based on the actions performed by the current policy [13].

DQN (Deep Q Network): Using the neural network model, the DQN calculates the Q-function. For calculation it uses the present value of Q as an input, so the output is the Q-value of each task [13].

Transductive Methods: In transductive methods, there is no difference between the training phase and the testing phase except that they use the prediction input of the training data. This type of method does not build a classifier for the whole input [9].

Self-Training: At the start of the self-training method, a classifier based on supervised learning is formed on a trained label set of data and the remaining data is used to gain predictions for an unlabeled set of data. Then, the resulting unlabeled data are added into the labeled data. After that, the supervised algorithm uses both of these data, which are labeled and unlabeled, to get the new version of data [14].

Figure 2.3 shows the comparative usage of popular ML algorithms in the medical domain. The data is generated by searching PubMed for machine learning algorithms in healthcare. It clearly indicates that the supervised ML algorithms – SVM and Artificial Neural Network are widely used in healthcare domain as compared to other ML algorithms.

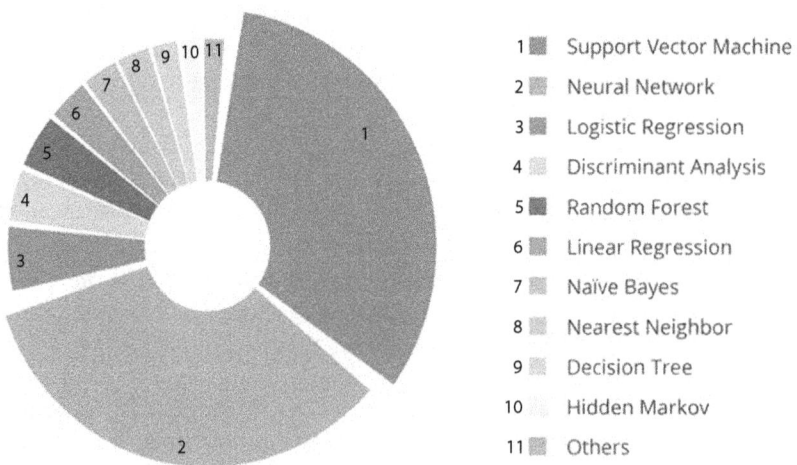

1	Support Vector Machine
2	Neural Network
3	Logistic Regression
4	Discriminant Analysis
5	Random Forest
6	Linear Regression
7	Naïve Bayes
8	Nearest Neighbor
9	Decision Tree
10	Hidden Markov
11	Others

FIGURE 2.3
Comparative usage of popular machine learning algorithms in the healthcare domain.

2.4 Disease Detection Using ML

Nowadays, machine learning is recognized as an expert way to diagnose various diseases because of its ability to give instant and precise results. Many medical practitioners can take the decisions in the early stages of a disease before it leads to any critical condition. In this section, we discuss some common diseases which can be diagnosed and treated in the early stages using the machine learning methods and algorithms that we have discussed in earlier sections.

2.4.1 Thyroid Disease

Survey [15] states that ten out of hundred humans are suffering from thyroid disease. Thyroid glands regulate essential body functions like inhalation, body weight, breathing and also the strength of the muscles. The two active hormones of thyroid, namely, thyroxin and triiodothyronine affect creation of protein and energy, body temperature and regulation of the human body. If the thyroid gland gets infected, then the human body loses essential control and, thus, may be dangerous for many patients. The thyroid can be the root cause behind various body disorders such as increase in sugar levels, cholesterol levels, low fertility and mental health among others. Four major categories of thyroid disorder are: hyperthyroid, hypothyroid, euthyroid and sick euthyroid.

2.4.1.1 Methodology

Figure 2.4 shows the architecture of thyroid prediction where different analyzing algorithms can be used for prediction. Different researchers implement different analyzing algorithms; following are the most commonly used methods in thyroid disorder detection:

SVM: The SVM method gives 90–93% accuracy on given data, using the recursive elimination function, so it helps doctors to differentiate patients into four classes and plan their treatment accordingly [16]. As discussed in Section 2.3, the SVM method uses the hyper plane concept for making decisions [17]. Most researchers use the SVM as a thyroid detection technique as it gives better results on any given data compared with other techniques.

Logistic Regression Classification: The logistic regression classification classifies the data based on the sigmoid function. On evaluating the logistic regression classifier on this thyroid data set, it shows a validation misclassification percentage of 18.76% and test misclassification percentage of 15.6% [18]. Using this logistic regression, we can get 80–85% accuracy on data.

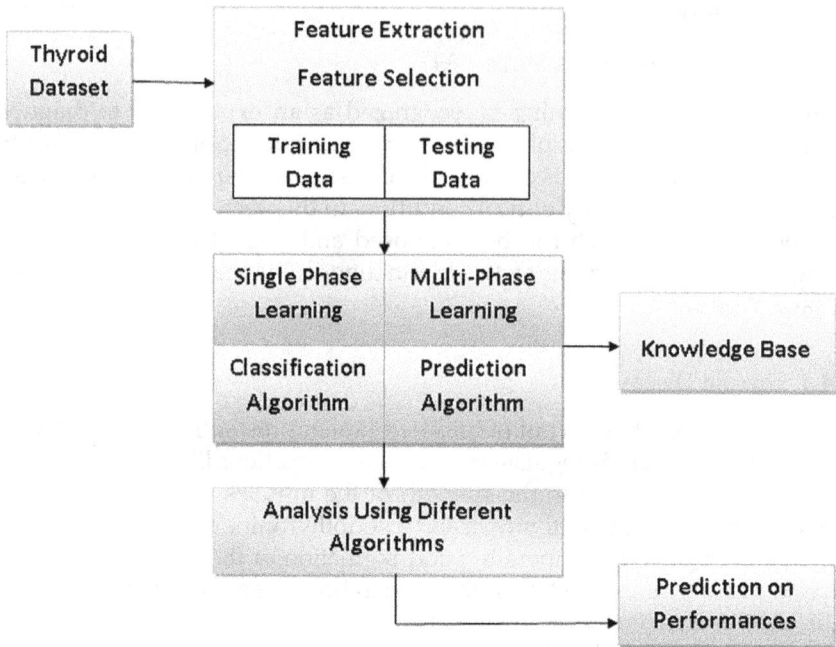

FIGURE 2.4
Architecture of thyroid prediction [4].

As an end result we can state that the SVM gives accuracy levels superior to any other algorithm in thyroid detection.

2.4.2 Heart Disease

Heart disease is one of the most significant causes of mortality in the world today. Prediction of cardiovascular disease is a critical challenge in the area of clinical data analysis. Identifying heart disease is very difficult because of other threat factors to the human body like cholesterol, sugar, and blood pressure levels. The heart controls the flow of the blood in the blood vessels. This provides oxygen and different types of nutrients to the body and also removes the waste through the blood. Lack of blood in the body is life threatening because the patient can slip into a critical condition. For the detection of heart disease, different techniques like ANN and data mining, among others, can be implemented [19]. Table 2.2 shows heart disease hazard factors and various symptoms that are to be considered in the detection of heart attacks.

TABLE 2.2

Heart Disease Causative Factors and Their Symptoms [16]

Heart Disease Hazard Factor	Heart Attack Symptoms
High cholesterol	Shortness of breath
High blood pressure	Pain in chest and anxiety
Diabetics	Ice-cold sweat and instability
Consuming too much alcohol	Uneven or fast heart beats
Being overweight	Anomalous pain
Smoking	Coronary vein disease

2.4.2.1 Methodology

The following different methodologies are useful in the detection and diagnosis of heart disease:

Naïve Bayes: The Naïve Bayes method differentiates the data using the Bayes theorem. The Bayes theorem uses the probability function to calculate the data. The Bayes function requires self-governing postulation and self-sufficient variables [21]. The accuracy level of the Naïve Bayes method in heart disease prediction is 85–86% [22].

Decision Tree Algorithm: The DT algorithm is used to create a training model in the form of a tree structure, which is based on preceding data, to calculate the class or target variables of novel data with the help of a DT algorithm. The accuracy level of prediction of this algorithm is superior to the other algorithms. The reason behind the higher accuracy is its analysis of each and every node during calculations. As discussed in Section 2.3, DT can analyze the data using three nodes – root, internal and leaf nodes [21]. The DT model predicts heart disease in patients with a precision level of 90–92% [22].

Thus, we conclude that the DT algorithm is the best choice for handling data regarding heart disease detection. In addition, it gives better results than the Naïve Bayes algorithm.

2.4.3 Breast Cancer

Breast cancer nowadays occurs very commonly among women and is becoming one of the most common causes of death in women. It has been proven that early detection and proper treatment can significantly increase the rate of survival. Mammography is a common test for diagnosing breast cancer. Mammograms are the films produced by a radiologist with the help of a specialized device. Using mammogram reports, doctors can diagnose the stage of the disease and can start treatment immediately. Common limits in breast cancer are tumors. A tumor can be one of two types: non-cancerous (benign), or cancerous (malignant), and can be detected in self-examination [23].

2.4.3.1 Methodology

The breast cancer system comprises four parts: (1) preprocessing, (2) features abstraction, (3) classification, and (4) data training and data testing. Using methods such as the support vector machine (SVM), Decision Tree (DT), or Naïve Bayes, the report can be segregated into normal or cancerous images. Then, extracted input is passed to ML algorithms like the support vector machine, Gaussian method and so forth [24].

Most research papers consist of two main parts: (1) prediction models and (2) preprocessing of data. Researchers have used the random forest algorithm, Naive Bayes algorithm, and SVM, among others, for prediction. Khuriwal et al. represented the breast cancer model [25]. Many researchers now use the breast cancer prediction model as shown in Figure 2.5. The breast cancer model consists of label encoder, normalizer and standard

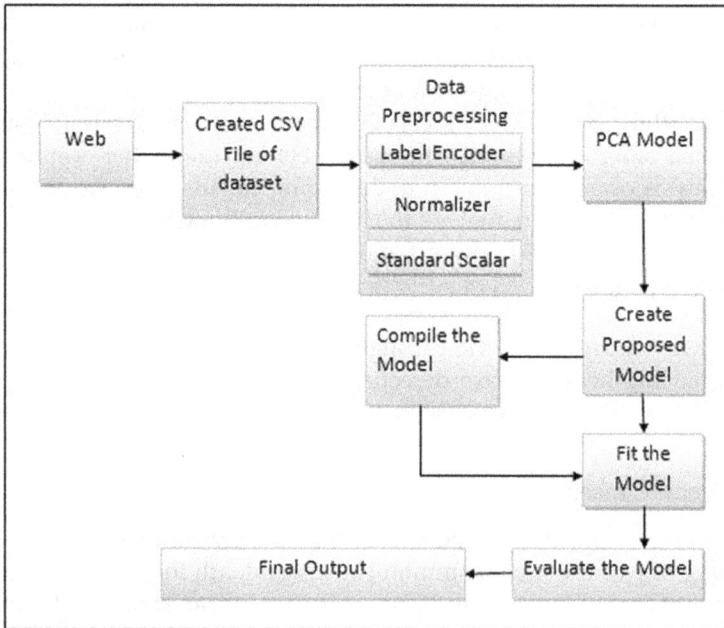

FIGURE 2.5
Breast cancer model [19].

scalar algorithm in the preprocessing stage for encoding the unconditional features into number values.

Label Encoder is a well-organized tool for encoding the levels of classes of features into numeric data. Data normalization is the process of rescaling one or more attributes in a range from 0 to 1. Principal Component analysis is used to maximize the conflict subsequently to the axis. In most papers, authors have used this model mainly with a deviation of 1 [26]. Using the breast cancer model clinical experts can detect breast cancer in early stages.

2.4.4 Diabetes

When blood contains increasing levels of sugar due to the confrontation of producing insulin in human body, it is called diabetes. It affects various organs in the body such as the kidneys, nerves and eyes if it is not promptly diagnosed [6]. For early detection of diabetes various ML algorithms can used, such as:

- Logistic Regression
- Naïve Bayes
- Stochastic Gradient Descent
- K-Nearest Neighbors
- Decision Tree
- Random Forest
- Support Vector Machine [26].

2.4.4.1 Methodology

Various ML techniques are extremely efficient methods for the early detection of diabetes as they have great classification ability. Some methods are discussed here [21]:

Adaboost: This is the first successful algorithm in diabetes detection and was developed for classification in binary form. It highlights classification problems and converts weak classification into stronger forms. This algorithm is based on the Decision Tree because the DT uses a tree structure and as trees are small in shape it simplifies decision making [26].

Bagging: The major purposes of bagging are for types of classification and regression techniques [27].

Random Forest (RF): Random forest is also used widely in classification and regression techniques [6]. By using a given sample, a random sample of the trained trees is created. It gives an accuracy level of 86–87% [36].

Thus, we conclude that for diabetes prediction and detection, the random forest (RF) algorithm gives better results when compared with the others.

2.4.5 Voice Disorder

Voice disorders are often encountered in patients like children, teachers and singers and even in normal people after some serious health problems, but is often neglected as it is not a life-threatening disease. Voice disorders can be detected and rectified using ML where some neural features are extracted from voice recordings. Voice recordings can be divided into two parts such as normal and pathological voice. The results show the common feasibility of deep learning and feature learning for the automatic recognition of voice disorders [28].

2.4.5.1 Methodology

The Convolutional Neural Network (CNN) is the most widely used deep learning technique. It is composed of many convolution layers and pooling layers and is popular for speech recognition and instrument detection using image classification. These images are passed to the Melspectrogram, so input can go through multiple layers for result calculation. The output received from the Softmax function, which can forecast the probabilities of all groups for a data point so this data point, will be assigned to the class with the larger possibility. The CNN architecture contains two convolutional layers: (1) pooling layer and (2) multiple layers. In a CNN the layers are designed for extraction of features and compared to the raw input representation, the learned representation is expected to be more representative. After output has been collected from the layers, it is passed to a Support Vector Machine SVM [28]. Each input layer ensures that the first layer and last layer are processed by a suitable number of bands. So, the input and convolutional layers are the same in each band. The top layers of CNN are combined with dissimilar structures extracted from the lower layers for the end result [29].

Figure 2.6, represents the CNN architecture. In this, the two layers used are max pooling and convolutional filters. Using these, we can detect voice disorders at an early stage and diagnose patients before they reach a critical condition.

In this section we have discussed different ML techniques and effective applications in the healthcare domain. We can draw some conclusions from the study and use of different ML/DL techniques. It is observed that the effectiveness of the method depends on the type and volume of input data. Table 2.3 summarizes effective techniques used for the detection of various diseases.

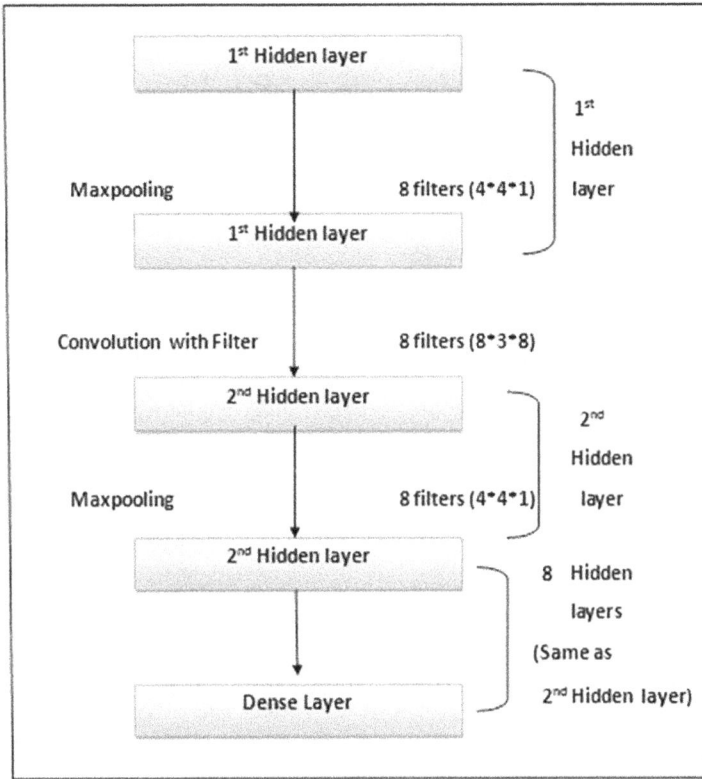

FIGURE 2.6
CNN architecture [32].

TABLE 2.3

ML Technique/Methodology Used in Detection of Disease

Sr. No.	Detection of Disease	ML Technique/ Methodology	Pros	Cons
1	Thyroid	SVM [16] Logistic regression classification [18]	SVM method gives 90–93% accuracy on given data using the recursive elimination function so it helps doctors to differentiate the patients into the four classes and plan their treatment accordingly.	SVM requires higher training time and hence, does not work on larger datasets. Does not perform well with overlapping classes.

(Continued)

TABLE 2.3   (Continued)

Sr. No.	Detection of Disease	ML Technique/ Methodology	Pros	Cons
2	Heart Disease	Decision Tree, Naïve Bayes [22]	The Bayes theorem uses the probability function to calculate the data. DT analyzes each and every node during calculation.	DT takes time to calculate each and every node.
3	Breast Cancer	SVM, Naïve Bayes [25] Decision Tree [24]	SVM & Naïve Bayes methods are well-organized tools for encoding the levels of classes of features into numeric data.	Using these methods always requires filter data for processing.
4	Diabetes	Decision Tree (DT) [27] Random Forest (RF) [26]	DT and RF highlight classification problems and convert the weak classification into the stronger.	Converting weaker to stronger RF requires the exact data.
5	Voice Disorders	CNN, SVM [27]	CNN and SVM show the common feasibility of deep learning and feature learning for automatic recognition of voice disorders.	These approaches have high dependency on the training data.

2.5 Security in Machine Learning/Deep Learning (ML/DL) for Healthcare

In this section, we investigate some major security issues of ML models in the healthcare domain and discuss various associated challenges [29]. Healthcare systems possess sensitive data related to patients' vital signs, and hence, have to confront various causes of vulnerabilities at all of these stages. Different types of security challenges at each stage of the pipeline for data-driven predictive clinical care are shown in Figure 2.7 [30].

Vulnerabilities in Data Collection: Clinical decision support needs a large amount of data in the form of patients' electronic health records (EHRs), types of medical imagery and radiology reports and many more. The

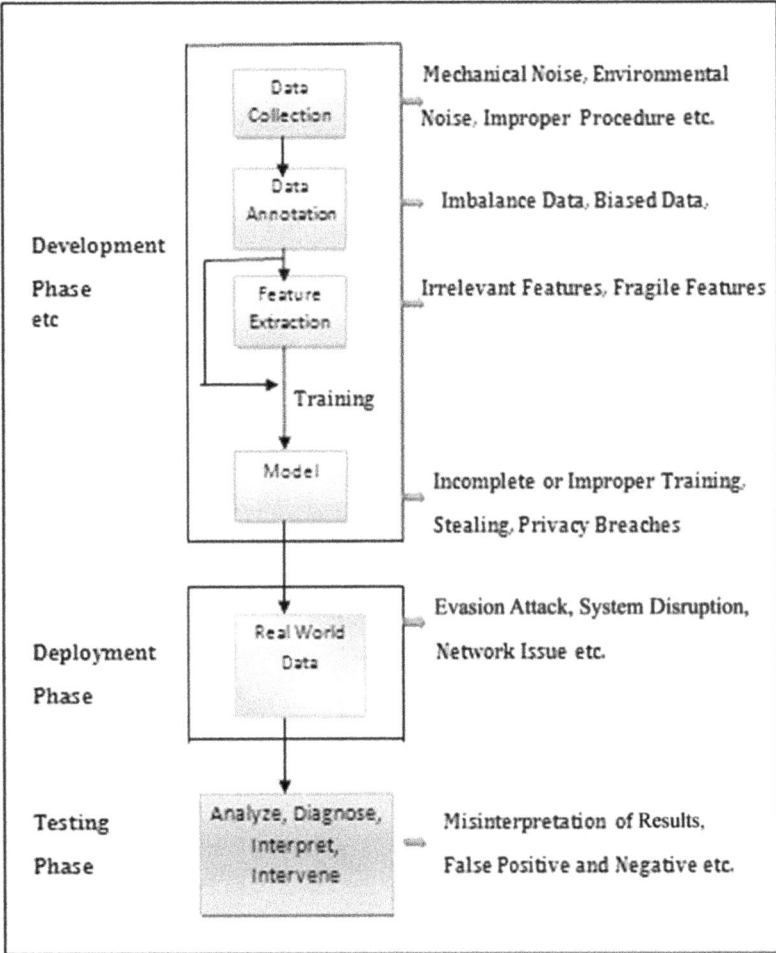

FIGURE 2.7
Stages of data-driven clinical care predictions and sources of vulnerabilities at all the stages [30].

collection and management of data is absolutely essential and time consuming. Collecting, storing and handling of this data requires utmost care as it plays a crucial role in diagnosis and further treatment. It is highly sensitive as it contains patient-specific personal healthcare records. There are many opportunities for, and sources of, vulnerabilities such as environmental and instrumental noise, human error while collecting data because of untrained clinical staff and some wrong procedures of data collection, which directly affect the functionality and efficiency of the ML system [31, 33].

Vulnerabilities Due to Data Annotation: Supervised learning is preferable for use in most applications of ML in healthcare systems. It requires a huge amount of training data to achieve high accuracy and precision for correct diagnosis. In supervised learning, in the data training phase, data samples need to be assigned specific labels, in a process called 'annotation'. It must be done correctly for proper results. This is the stage where different vulnerabilities may occur, such as data imbalance, data biasing and misspecification or wrong data labeling [34].

Vulnerabilities in Model Training: Model training is the most important stage in the ML system. The efficacy of decisions made by the system depend solely on how the system is trained. For this training, some important training parameters, such as learning rate, epochs and batch size must be used correctly. At this stage, many vulnerabilities may occur such as wrong or incomplete training, ML model poisoning and some privacy breaches due to improper selection and use of these training parameters [35].

2.6 Security of ML

We discussed types of vulnerabilities in the previous section. Here, we provide an outline of ML security predominantly from the viewpoint of the healthcare system and will further discuss the security challenges in ML. Healthcare systems store and maintain very confidential and important data records of patients' health, which are to be protected essentially from increased risk factors like malicious attacks, malware, data breaches and viruses. Different data security measures are an essential part of the healthcare industry and protect this subtle and highly sensitive information [36]. At the same time, the ML engines which process abundant data need constant monitoring of both the network and other system parts for immediate detection any kind of threat or attack.

2.6.1 Security Threats

As shown in Figure 2.8, there are three broad categories of security threats on ML systems namely: (1) influence attacks, (2) security violations, and (3) the attack's specificity [37].

1. *Influence Attacks*: Following are the two types of influence attacks:
   a. Causative: This type of attack investigates to gain control over training data.
   b. Exploratory: This type of attack exploits incorrect classification of the ML model without overriding the model training [27].

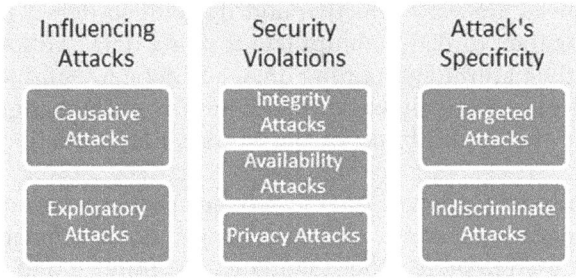

FIGURE 2.8
A categorization of security threats in ML [35].

2. *Security Violations*: Security is concerned with two issues – the availability and integrity of the services. There are three types of security violations:

   a. Integrity Attack: This tries to increase the false-negative rate of the deployed model (classifier) when the model is given harmful inputs.

   b. Availability Attack: Contrasting to an integrity attack, this tries to achieve an increase in the false-positive rate of the classifier in response to benign inputs.

   c. Privacy Violation Attack: This is related to the introduction of subtle and confidential information of the training data and trained model or both [27].

3. *Attack's Specificity*: There are two ways of defining specificity attacks:

   a. Targeted attack: This type of attack is proposed for some specific input samples or a group of samples.

   b. Indiscriminate attack: This causes the ML model to miscarry comprehensively [27].

Adversarial ML attacks have resulted from current hard work in the identification of vulnerabilities in ML model training and conclusion. Adversarial attacks are considered to be one of the major security threats to ML systems [37]. In general, there are two types of adversarial attacks:

- *Poisoning Attacks*: These attacks manipulate the training data while training the model. That is, it misleads the training of ML.

- *Evasion Attacks*: These attacks are activated during the conclusion phase of the learning or training stage [37]. The testing data is influenced by attackers which destroys the integrity of the ML model due to harmful inputs.

Poisoning attacks are very effective and difficult to detect in healthcare applications as they directly manipulate training data. Although it is difficult to directly manipulate training data, adding new samples into training data might be relatively easier task. But any such change in training data obstructs the applicability of the ML system in healthcare applications [38].

In ML security and privacy, AML is a major problem for healthcare and biometric applications and can cause major unintentional conditions. Biometrics are used to protect the systems against fraud and to protect confidential patient records and for added security in medical facilities and equipment [39].

2.7 Conclusion and Future Scope

Health is the topmost priority of any human being and hence, in today's economy, healthcare is one of the fastest growing segments. Technologies such as big data, AI and ML have the potential to help both patients and healthcare providers by offering better services at lower costs. Many organizations have already started servicing the healthcare industry and have helped to facilitate disease detection and prediction for patients' correct treatment at earlier stages.

In this chapter we have discussed different diseases where ML can help to detect and predict the growth of the disease in early stages. We need to collect the data in various formats and need to interpret it using ML methods like the Decision Tree, Adaboost, Random Forest and SVM. These techniques are useful to diagnose the disease promptly, enabling doctors to take decisions as soon as possible and start treatment, so that patients may recover from the disease more quickly. Thus, machine learning is very useful in the healthcare sector. Table 2.4 summarizes a survey to find the most effective ML techniques used for specific disease detection and prediction.

In future, there will be different ways in which machine learning techniques can be applied for effective disease prediction, diagnosis and treatment, improving overall operations in the healthcare domain. This survey will help researchers in disease detection and prediction using more effective ML techniques. The increasing number of applications of machine learning in healthcare allows us to glimpse a future where data, analysis and innovation work hand-in-hand to help countless patients without them ever realizing it.

TABLE 2.4

Summary

Sr. No.	Disease Name	Most Effective ML Method
1	Thyroid	Use of the Random Forest algorithm gives the best results in thyroid detection, because it identifies malignancy, especially for abnormal nodules, in terms of risk score.
2	Breast cancer	In breast cancer detection, the SVM algorithm gives better results than any other algorithm because SVM has an extra advantage of automatic model selection, in the sense that both the optimal number and locations of the basic functions are automatically obtained during training.
3	Heart disease	The Decision Tree algorithm provides the highest accuracy of all methods in heart pulses, so recognizing heart disease in earlier stages is possible, allowing patients to get treatment as early as possible.
4	Diabetes	The Adaboost and bagging algorithm gives the best results in diabetes detection especially with accuracy for diabetes and control satisfaction data.
5	Voice disorders	The Convolution Neural Network (CNN) is used as an effective deep learning method and hence is recognized as the best method used for detection of voice disorders. Using CNN we can also detect abnormalities in the singing voice, so it is best for voice disorder.

References

[1] Singh, A., & Kumar, R. (2020, February). Heart Disease Prediction Using Machine Learning Algorithms. *2020 International Conference on Electrical and Electronics Engineering (ICE3)* (pp. 452–457). IEEE.

[2] Shailaja, K., Seetharamulu, B., & Jabbar, M. A. (2018, March). Machine Learning in Healthcare: A review. *2018 Second International Conference on Electronics, Communication and Aerospace Technology (ICECA)* (pp. 910–914). IEEE.

[3] Berry, M., Potok, T. E., Balaprakash, P., Hoffmann, H., & Vatsavai, R. (2015). *Machine Learning and Understanding for Intelligent Extreme Scale Scientific Computing and Discovery. DOE Workshop Report, January 7–9, 2015, Rockville, MD*. USDOE Office of Science (SC), Washington, DC (United States). Advanced Scientific Computing Research (ASCR).

[4] Nagarhalli, T. P., Save, A. M., & Shekokar, N. M. (2021). Fundamental Models in Machine Learning and Deep Learning. In *Design of Intelligent Applications Using Machine Learning and Deep Learning Techniques* (pp. 13–36). Chapman and Hall/ CRC.

[5] Tyagi, A., Mehra, R., & Saxena, A. (2018). Interactive Thyroid Disease Prediction System Using Machine Learning Technique. *2018 Fifth International Conference on Parallel, Distributed and Grid Computing (PDGC)*.

[6] Ma, L., Ma, C., Liu, Y., & Wang, X. (2019). Thyroid Diagnosis from SPECT Images Using Convolutional Neural Network with Optimization. *Computational Intelligence and Neuroscience*, 2019, pp. 1–11.

[7] Ray, S. (2019, February). A Quick Review of Machine Learning Algorithms. *2019 International Conference on Machine Learning, Big Data, Cloud and Parallel Computing (COMITCon)* (pp. 35–39). IEEE.

[8] Amruthnath, N., & Gupta, T. (2018). A Research Study on Unsupervised Machine Learning Algorithms for Early Fault Detection in Predictive Maintenance. *2018 5th International Conferenceon Industrial Engineering and Applications (ICIEA)*. doi:10.1109/iea.2018.8387124

[9] Van Engelen, J. E., & Hoos, H. H. (2020). A Survey on Semi-Supervised Learning. *Machine Learning*, vol. 109, no. 2, pp. 373–440.

[10] Ogbuabor, G., & Ugwoke, F. N. (2018, April). Clustering Algorithm for a Healthcare Dataset Using Silhouette Score Value. *International Journal of Computer Science & Information Technology (IJCSIT)*, vol. 10, no. 2, pp. 27–37.

[11] Yu, C., Liu, J., Nemati, S., & Yin, G. (2021). Reinforcement Learning in Healthcare: A Survey. *ACM Computing Surveys (CSUR)*, vol. 55, no. 1, pp. 1–36.

[12] Collins, A., & Yao, Y. (2018). Machine Learning Approaches: Data Integration for Disease Prediction and Prognosis. *Applied Computational Genomics*. Springer, pp. 137–141.

[13] Samsuden, M. A., Diah, N. M., & Rahman, N. A. (2019). A Review Paper on Implementing Reinforcement Learning Technique in Optimising Games Performance. *2019 IEEE 9th International Conference on System Engineering and Technology (ICSET)*.

[14] Choudhary, G., & Narayan Singh, S. (2020). Prediction of Heart Disease using Machine Learning Algorithms. *2020 International Conference on Smart Technologies in Computing, Electrical and Electronics (ICSTCEE)*.

[15] Chaubey, Gyanendra, Bisen, Dhananjay, Arjaria, Siddharth, & Yadav, Vibhash. (2020). Thyroid Disease Prediction Using Machine Learning Approaches. *National Academy Science Letters*, vol. 44. doi:10.1007/s40009-020-00979-z

[16] Duggal, P., & Shukla, S. (2020). Prediction of Thyroid Disorders Using Advanced Machine Learning Techniques. *2020 10th International Conference on Cloud Computing*, Data Science & Engineering (Confluence).

[17] Wang, Q., Cao, W., Guo, J., Ren, J., Cheng, Y., & Davis, D. N. (2019). DMP_MI: An Effective Diabetes Mellitus Classification Algorithmon Imbalanced Data with Missing Values. *IEEE Access*, vol. 7, pp. 102232–102238.

[18] Chubey, G., & Bisen, D. (2020). Thyroid Disease Prediction Using Machine Learning Approaches. The National Academy of Sciences, India 2020.

[19] Mohan, S., Thirumalai, C., & Srivastava, G. (2019). Effective Heart Disease Prediction Using Hybrid Machine Learning Techniques. *IEEE access*, 7, 81542–81554.

[20] Obulesu, O., Mahendra, M., & ThrilokReddy, M. (2018, July). Machine Learning Techniques and Tools: A Survey. In *2018 International Conference On Inventive Research In Computing Applications (ICIRCA)* (pp. 605–611). IEEE.

[21] Tripathi, G., & Kumar, R. (2020). Early Prediction of Diabetes Mellitus Using Machine Learning. In *2020 8th International Conference on Reliability, Infocom Technologies and Optimization (Trends and Future Directions) (ICRITO)*.

[22] Singh, A., & Kumar, R. (2020). Heart Disease Prediction Using Machine Learning Algorithms. *2020 International Conference on Electrical and Electronics Engineering (ICE3-2020).*

[23] Kajala, A., & Jain, V. K. (2020). Diagnosis of Breast Cancer Using Machine Learning Algorithms – A Review. *2020 International Conference on Emerging Trends in Communication, Control and Computing (ICONC3).*

[24] Hussain, L., Aziz, W., Saeed, S., Rathore, S., & Rafique, M. (2018). Automated Breast Cancer Detection Using Machine Learning Techniques by Extracting Different Feature Extracting Strategies. *2018 17th IEEE International Conference on Trust, Security and Privacy in Computing and Communications/12th IEEE International Conference on Big Data Science and Engineering (TrustCom/BigDataSE).*

[25] Khuriwal, N., & Mishra, N. (2018). Breast Cancer Diagnosis Using Deep Learning Algorithm. *2018 International Conference on Advances in Computing, Communication Control and Networking (ICACCCT).*

[26] Islam, M. T., Raihan, M., Aktar, N., Alam, M. S., Ema, R. R., & Islam, T. (2020). Diabetes Mellitus Prediction Using Different Ensemble Machine Learning Approaches. *2020 11th International Conference on Computing, Communication and Networking Technologies (ICCCNT).*

[27] Learning Strategies for Voice Disorder Detection. Hongzhao Guan, Center for Music Technology, Georgia Institute of Technology, Atlanta, Georgia 30332. Email: hguan7@gatech.edu. Alexander Lerch, Center for Music Technology, Georgia Institute of Technology, Atlanta, Georgia 30332 (2019).

[28] Qayyum, A., Qadir, J., Bilal, M., & Al Fuqaha, A. (2020). Secure and Robust Machine Learning for Healthcare: A Survey. *IEEE Reviews in Biomedical Engineering*, pp. 1–1.

[29] Hinton, G. E., & Salakhutdinov, R. R. (2006). Reducing the Dimensionality of Data with Neural Networks. *Science*, vol. 313, no. 5786, pp. 504–507.

[30] Karimian, N., Tehranipoor, M., Woodard, D., & Forte, D. (2019). Unlock Your Heart: Next Generation Biometric in Resource-Constrained Healthcare Systems and IoT. *IEEE Access*, vol. 7, pp. 135–149.

[31] Caruana, R., Lou, Y., Gehrke, J., Koch, P., Sturm, M., & Elhadad, N. 2015. Intelligible Models for Healthcare: Predicting Pneumonia Risk and Hospital 30-Day Readmission. *Proceedings of the 5th ACM SIGKDD International Conference on Knowledge Discovery and Data Mining*. ACM, pp. 175–1730.

[32] Bhagwat, Harshita, & Gogate, Uttara. (2021). Learning CNN Strategy for Voice Disorder Classification and Detection. *IJCRT* vol. 9, no 7, July 205, pp. a844–a851. ISSN: 2320-2882.

[33] Shekokar, N. M., Shah, C., Mahajan, M., & Rachh, S. (2015). An Ideal Approach for Detection and Prevention of Phishing Attacks. *Procedia Computer Science*, vol. 49, pp. 82–91.

[34] Szegedy, C., Zaremba, W., Sutskever, I., Bruna, J., Erhan, D., Goodfellow, I., & Fergus, R. (2013). Intriguing Properties of Neural Networks. arXiv preprint arXiv:1312.6199

[35] Fredrikson, M., Jha, S., & Ristenpart, T. (2015). Model Inversion Attacks That Exploit Confidence Information and Basic Countermeasures. *Proceedings of the 22nd ACM SIGSAC Conference on Computer and Communications Security*. ACM, pp. 1322–1333.

[36] Potluri, S., Mangla, M., Satpathy, S., & Mohanty, S. N. (2020, July). Detection and Prevention Mechanisms for DDoS Attack in Cloud Computing Environment. *2020 11th International Conference on Computing, Communication and Networking Technologies (ICCCNT)*. IEEE, pp. 1–6.

[37] Papernot, N., McDaniel, P., Jha, S., Fredrikson, M., Celik, Z. B., & Swami, A. (2016). The Limitations of Deep Learning in Adversarial Settings. *2016 IEEE European Symposium on Security and Privacy (EuroS&P)*. IEEE, pp. 372–387.

[38] Mozaffari-Kermani, M., Sur-Kolay, S., Raghunathan, A., & Jha, N. K. (2014). Systematic Poisoning Attacks on and Defenses for Machine Learning in Healthcare. *IEEE Journal of Biomedical and Health Informatics*, vol. 19, no. 6, pp. 1893–1905.

[39] Papangelou, K., Sechidis, K., Weatherall, J., & Brown, G. (2018). Toward an Understanding of Adversarial Examples in Clinical Trials. *Joint European Conference on Machine Learning and Knowledge Discovery in Databases*. Springer, pp. 35–51.

[40] Finlayson, S. G., Bowers, J. D., Ito, J., Zittrain, J. L., Beam, A. L., & Kohane, I. S. (2019). Adversarial Attacks on Medical Machine Learning. *Science*, vol. 363, no. 6433, pp. 1287–1289.

3

Scope of Machine Learning and Blockchain in Cyber Security

Rohini Patil, Monika Mangla and Smita Bansod

University of Mumbai, Mumbai, India

CONTENTS

3.1 Introduction

There has been a tremendous technological transformation during the past few decades. This transformation has left hardly any aspect of human life untouched. The main aspects of technological revolution can be considered as high internet connectivity leading to a sophisticated lifestyle, ease in communication and convenience of service. However, this advancement also has some associated challenges in terms of security flaws. These security flaws may lead to serious consequences in terms of finance, reputation and much more if proper security measures are not adopted. Consequently, there has been an increase in cybercrime during recent years, which could be considered an associated challenge. Cybercrime is not limited to any specific area but has impacted different sectors like the healthcare, financial and educational sectors among others [1]. Hence, considering the wide range of cybercrimes, it becomes necessary to understand the associated threats and challenges so as to manage them more effectively.

DOI: 10.1201/9781003408307-4

TABLE 3.1

Various Types of Attacks and Corresponding Occurrence Percentages [2]

Type of Attack	Percentage (%)
Malware	32.3
Unknown	21.7
Account Takeover	14.2
Vulnerability	13.2
Targeted Attack	5.5
Misconfiguration	2.9
Fake Website/Social Network Accounts	1.7
Business Email Compromise	1.4
DDoS	1.4
Others	5.7

We have witnessed an exponential rise in cybercrime owing to the rapid expansion of cyberspace obfuscating the overall impact. As a result, ensuring cyber security has become an extremely complex task and requires domain knowledge in terms of analyzing the possible vulnerabilities. To this end, different techniques have been employed such as artificial intelligence (AI), machine learning (ML) and advanced data mining. Among various approaches, ML has been widely employed in cyber security techniques such as user authentication, intrusion detection system (IDS) and malware detection among others. At the same time, blockchain is gaining unprecedented acceptance as a technology to develop secured application. Following a survey of literature, it is observed that several researchers have advocated implementation of blockchain to prevent cyber-attacks based on their research and experience.

The sole objective of the cyber-attack is to gain access to the system of the victim so that the attacker can perform any unauthorized operations. During the pandemic outbreak, there was an exponential rise in cyber-attacks owing to increased communication over networks. Worldwide statistics for cyber-attacks that took place during April 2021 are shown in Table 3.1.

As stated earlier, in order to address the network vulnerabilities, it is imperative to understand the possible vulnerabilities and hence, the authors present the various types of web-based attacks as follows:

1. **SQL Injection**: A Structured Query Language (SQL) injection is a web-based attack. In this attack, an attacker inserts malicious code (preferably in the form of a query) and forces the exposure of sensitive data from the database.

2. **Cross Site Scripting Attack**: In these attacks, attackers insert some malicious code into the webpage with the intention to steal the

cookies of users. As a result, the malicious script gives the attacker cookies for the session, such as log keystrokes.

3. **Malware**: This is malicious software intended to destroy a system or a network. Elementary software includes viruses, bot, worms, Trojans, spyware and ransomware. Generally, malware infects the system by users clicking on a hazardous link of an email attachment or a hazardous link appearing during software installation that leads to stealing sensitive large-scale data and the installation of additional spiteful software.

4. **Phishing**: Phishing is fraudulent email communication intending to steal sensitive personal information like login credentials or credit card information, or to install malware on the victim's system.

5. **Man-in-the-middle**: When attackers insert themselves into a two-party transaction, eavesdropping on the entire communication without the knowledge of end parties, it is termed as a man-in-the-middle attack. In such scenarios, all information passes through the attacker.

6. **Denial-of-service**: In these attacks, an entire network is flooded by sending an infinite number of host requests. Launching a flood of attacks slows down the user's network leading to a distributed-denial-of-service attack.

7. **Zero-day exploit**: This type of attack hits after announcing network vulnerability but ahead of implementing the corresponding solution.

8. **Cryptojacking**: This is a special type of attack where the attacker uses someone else's computer for producing cryptocurrency for the target. To perform the necessary calculations, the attackers will install malware on the victim's computer.

Other than these, some recent cyber-attacks include: AdvisorsBot, Andromeda, Cerber, Cryptoloot CNRig, HiddenMiner, Fireball, Nivdort, Iotroop, RubyMiner, NotPetya, Trickbot, AdultSwine, WannaCry, and cryptocurrency attacks and others that may cause severe damage [3].

In considering the wide range of network vulnerabilities in the cyber world, it is evident that cyber security is the most concerning aspect from the perspective of national and economic security policies. Recent challenges in the field of cyber security are presented in Figure 3.1 and then explained further.

- **Ransomware Evolution**: This type of malware locks the victim's computer data and may demand ransom for unlocking the data. It can target anyone including government offices, banks and small entities.

FIGURE 3.1
Challenges in cyber security.

- **Internet of Things (IoT) Threats**: The IoT system is interconnected with several physical devices accessible via the internet creating a room for imminent security flaws.
- **Artificial Intelligence Expansion**: AI builds an intelligent system capable of doing any work in a manner similar to humans. Nowadays there is a challenge if control of a machine is compromised by a breach in security.
- **Blockchain Revolution**: Blockchain technology is a vast global platform which enables cryptocurrencies so as to secure all associated security flaws.
- **Serverless Apps Vulnerability**: Serverless architecture and apps depend on third-party cloud infrastructure. If the serverless apps are not hosted with a server, cyber attackers easily spread threats. Therefore, this creates a challenging task while using serverless applications.

Now it is evident that there are several kinds of attacks that may happen in a network and hence, various preventive measures must be adopted to minimize network vulnerabilities. In order to measure the efficiency and effectiveness of a proposed approach, there are several performance metrics which have been widely accepted. Some of these performance measures are as follows:

- **Cost per example (CPE)**: This calculates the cost of the misclassification.

- **Transaction delay**: The time lapse between submission of a transaction and its completion.

- **Transaction throughput**: The number of transactions committed per second.

- **Consensus cost time**: The time taken for a transaction to be processed and validated.

- **Propagation delay**: The time taken to propagate a transaction throughput in the blockchain peer-to-peer network.

- **RPC response time**: The time required to complete a remote procedure call.

The supremacy of any proposed approach to achieve network security can be established by using these metrics.

The authors in this chapter aim to present the roles of machine learning and blockchain in controlling cyber-attacks. The chapter has been organized into various sections. First, Section 3.1 is dedicated to the role of ML and blockchain in cyber security. The scope of machine learning and blockchain in cyber security is discussed in Sections 3.2 and 3.3 respectively. The authors then propose a hybrid approach that integrates ML and blockchain in Section 3.4. Finally, the chapter is concluded in Section 3.5 by mentioning the future scope of research for potential researchers.

3.2 Machine Learning for Cyber Security

ML has an ability to learn without being explicitly programmed and works on the concept of mathematical modeling derived from patterns and the ability to generate predictions. It has been employed in numerous fields including e-commerce and the healthcare, financial and educational sectors, and many more. ML can be broadly categorized as supervised and unsupervised learning where supervised learning learns to predict using target variables while unsupervised learning is patterns-based learning, for example, identifying malwares with similar behavioral patterns.

Machine learning has the potential to give a promising performance in the field of cybercrime. Owing to advances in technology and research in related fields, it helps to identify and deal with cyber threats. Using machine learning, use cases for log analysis can be easily detected and mitigated. Some ML classification algorithms used in the security domain are: Decision Trees,

Support Vector Machines, Random Forest, K-Nearest Neighbor, ensemble learning, Bayesian algorithms and clustering algorithms [4].

3.2.1 Threat Model for ML

The threat model for ML, under adversarial settings, specifies the characteristics of the attacks and contains the three layers process as shown in Figure 3.2 [5].

1. **Defensive methods layer**: This layer takes defensive measures to protect from attacks by the following steps:
    *   Protection of the training data by isolating and rejecting the adversarial sample through data sanitization.
    *   ML classifiers are retrained including adversarial samples in order to improve robustness and security of the algorithms, so as to detect anomalies.
    *   In order to assess the security of ML classifiers, a risk assessment scheme is used.
    *   Several privacy mechanism and homomorphic encryption techniques are used to maintain the confidentiality and the privacy of the data [5].
2. **ML layer**: This layer contains steps such as data gathering, feature extraction, training and testing phases of classifier.
3. **Attack layer**: In this layer, an attacker can access the classifier through false data injection and stealthy channel attacks. This affects the integrity and availability of the models by injecting the adversarial samples during training. It uses the homomorphic scheme due to high protection and confidentiality. Also, the evasion attack breaches the security of model by modifying key features of the algorithm and gaining model authority. After deployment of the classifier, the attacker may exploit a stolen model by sending repeated queries. And finally, during the inference phase, attacks are classified into black-box attack and white-box depending upon the knowledge of the attacker model. Strong attackers launch white-box attacks, while weak attackers perform black-box attacks.

Various researchers have employed ML to prevent cyber-attacks and achieve cyber security. Here, authors review various cyber security threats and challenges in modern society and some proposed solutions to prevent these cyber threats, motivated by the work in [1]. The authors discuss the significance of cyber security in the current era in [3]. In the same context, authors

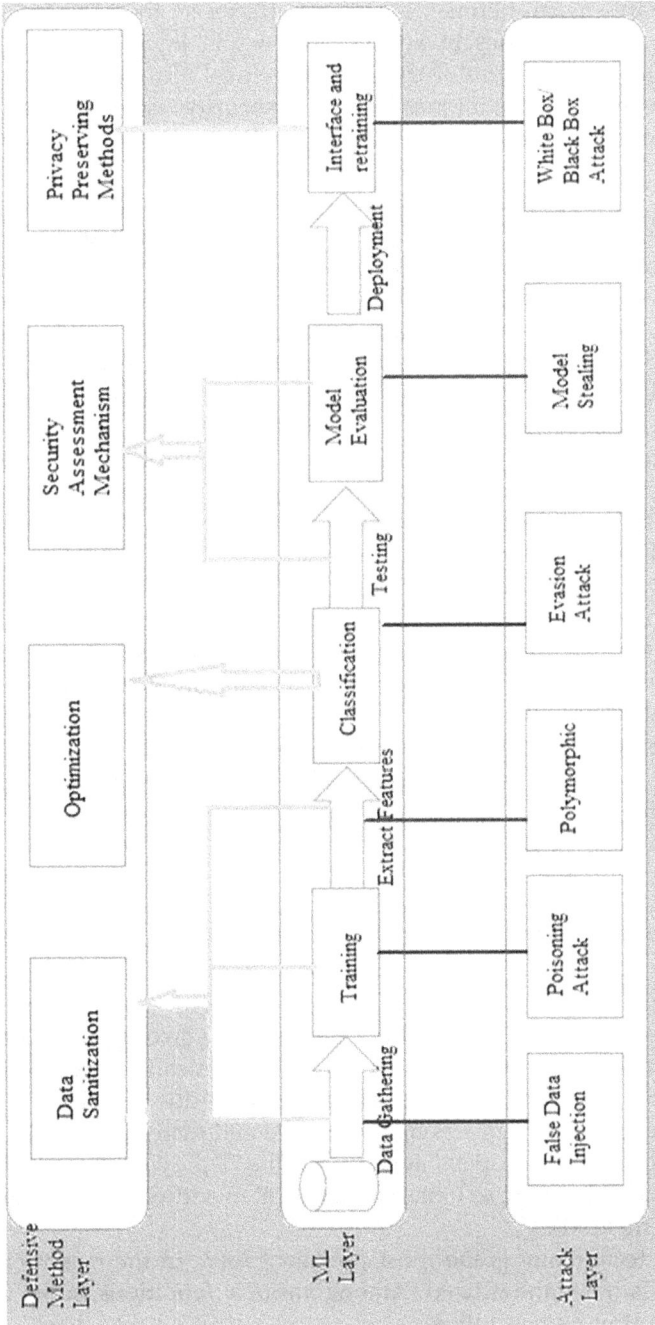

FIGURE 3.2
Threat model for ML process [5].

in [4] highlight the use of AI in cyber security. Authors in [5] propose a threat model for analyzing numerous adversarial attacks on the ML models and investigating the properties of attacks. Authors in [6] proposed a multi-layered framework to minimize the security issues. Authors in [7] also discuss an overview of AI techniques to resolve security issues and defending mechanisms against attacks, thus, establishing the competence of AI and ML in this field.

Authors in [8] propose a secure SaaS Framework for mitigation of attack that uses a Deep Belief Network (DBN) along with Median Fitness Oriented Sea Lion Optimization technique (MFSLnO) for attack detection. The model proposed in [8] yielded a throughput and packet ratio of 89% and 16% respectively. A similar problem is also tackled by authors in [9] by presenting a novel ML-based security framework related to the IoT domain. In this work, attack distribution using data mining helped to gain high performance and low-cost attack detection with an accuracy of 99.71%. Authors in [10] also review the challenges of ML techniques in terms of mobile and computer network attacks, considering different datasets, tools and evaluation metrics. Similarly, authors in [11] show recent and existing attacks and their effect using different mitigation approaches.

Authors in [12] propose an AI/ML-based hybrid model to handle security issues in the cloud network infrastructure for mitigating IoT cyber threats at network and host level. In order to predict response type and malware, authors in [13] developed a text-mining-based model that works on collected datasets. A comprehensive survey on use of ML in cyber security over five years starting from 2013 has been presented by authors in [14, 15] and encompasses the basics of cyber-attacks, defenses and algorithms.

3.3 Blockchain for Cyber Security

There are several traditional security approaches, however, these conventional approaches are computationally and energetically expensive. As a result, they are seldom applicable for the IoT scenario due to scalability issues [16]. As a result, the IoT requires light, portable and distributed privacy and security solutions, which opens avenues for the deployment of blockchain technology in this domain, as it has the potential to address the challenges of conventional methods.

Blockchain technology is the most prevalent topic in the recent era, but its definition is not standardized. Among various definitions, some widely accepted definitions are as follows:

- Distributed ledger with confirmed blocks organized in an append-only, sequential chain using cryptographic links [17] iso.org (iso:std:iso:22739)
- Tamper evident and tamper resistant digital ledgers implemented in a distributed fashion (i.e., without a central repository) and usually without a central authority (i.e., a bank, company, or government) [16] NIST (NISTIR 8202)
- A type of Distributed Ledger Technology (DLT) where transactions are recorded with an immutable cryptographic signature called a hash. The transactions are then grouped in blocks and each new block includes a hash of the previous one, chaining them together, hence why distributed ledgers are often called blockchains [18].

From the definitions of blockchain, it is clear that it is a distributed computing network that serves as a public ledger as well as a platform for secure transmission without the use of a third party. Blockchain is a technology that allows digital exchanging of money, just as the internet does. Any kind of data from currencies to land property or votes can be tokenized, stored and shared on a blockchain network. Blockchain technology was first manifested by Santoshi Nakamoto in 2009 with the Bitcoin application [19] as a peer-to-peer electronic cash system. Later, it gained tremendous popularity owing to its immunity to any interference by a third party. Currently, blockchain is not restricted to its application in bitcoin or cryptocurrency. Today, blockchain is evolving and it is observed in various applications and domains like government, the IoT, AI and so on.

Apart from numerous fields, blockchain has also emerged as a new weapon in cyber security [17]. The features of the blockchain are attractive in order to handle various security attacks as discussed above. The application of blockchain in cyber security is feasible owing to its feature of being a distributed ledger technology which is used to create trust between untrusted parties and works as a robust cyber security technology. Another exciting feature of blockchain is the security of private messages, which is a challenging task in this era of social media. For the same reason, blockchain can be used to form a unified API framework that enables cross-messenger communication capabilities. This is achieved owing to the InterPlanetary File System (IPFS) in blockchain that handles huge data with timestamp and secured metadata in blockchain. This IPFS structure of blockchain makes it nearly impossible for hackers to penetrate the data storage systems, thus, maintaining the data integrity.

A systematic literature review of blockchain in the field of cyber security is given in [20] where authors discuss various types of blockchain research areas as shown in Table 3.2.

TABLE 3.2

Blockchain Cyber Security Application Areas of Research [20]

Application Areas	Percentage (%)
IoT	45
Networks	10
Data Storage and Sharing	16
PKI	7
Data Privacy	7
Web	3
WiFi	3
DNS	6
Malware	3

According to a Gartner survey, IoT device consumption is expected to rise up to 25 billion which will eventually lure attackers to also try their hands at various security attacks that may lead to loss of finance or reputation. Blockchain comes to the rescue in such applications as it enables decentralization of the administration [21, 22]. Various renowned companies like Walmart, IBM, Microsoft and MasterCard, among others, have been utilizing blockchain for maintaining data security. Additionally, various IoT enabled companies have been adopting blockchain technology for enhanced security of their networks as it has proven its competence in handling various kind of security attacks.

As per the authors in [23, 24], blockchain has emerged as the best solution for DDoS attacks when combined with other methods. For instance, virtual artificial blockchain and deep learning and blockchain have the potential to provide defense against even DDoS attacks. The competence of blockchain-based solutions to mitigate DDoS attacks is also demonstrated by the authors in [25]. However, scalability and cost are found to be two prime concerns for the proposed model which need to be addressed.

The authors in [26] also present applications of blockchain in security-related use cases, by analyzing different aspects such as backup and recovery, threat intelligence and content delivery networks. Further, the usage of blockchain in reference to cyber-threats in Industry 4.0 has been discussed in [27]. In [27], the authors propose classification of cyber-attacks based on cover scanning, local to remote, power of root and denial of service (DoS) in Industry 4.0.

The employment of blockchain in cyber security has also been accepted by the authors in [28] as they propose a secure distributed model that facilitates cyber threat intelligence (CTI) sharing platform based on a private blockchain. The proposed model addresses generic CTI collaboration issues and enhances the defense against perspective threats, thus, mitigating the DDoS attacks.

The usage of blockchain has been integrated with other approaches to enhance the efficiency of secured models. For instance, the authors in [29] propose a deep autoencoder neural network which is supported by blockchain. The proposed model is used for classification and management of attack incidents. In the proposed model, authors use a blockchain-based smart contract technique that provides an automated trusted system and allows automatic acquisition, classification and enrichment of incident data. Further, it is demonstrated that the proposed technique can be applied to support incident handling tasks performed by security operation centers.

It can be drawn from the above discussion that blockchain brings major advantages to the security world as follows [30]:

1. **Tamper Proofing**: This is achieved by the unique data structure of blockchain and the corresponding writing mechanism. In blockchain, due to consensus algorithm, any writing operation requires approval of a specific percentage of users which is generally more than 50%. As a result, in order to perform any adverse operation, the attacker has to gain control over more than 50% of the nodes, and that requires stronger computing power, thus, making it tamper proof.

2. **Disaster Recovery**: In blockchain, each user keeps a full copy of the data. Although it leads to redundancy to some extent, this redundancy is bearable as it provides a reliable and fault tolerant network.

3. **Privacy Protection**: Blockchain uses asymmetric encryption mechanisms to allow users to use their own private keys to encrypt data. This private key has no dependence on the user's real identity, and hence, blockchain obtains data security while preserving user anonymity and privacy.

From the above discussion, we can observe that blockchain is gaining widespread acceptance as an efficient technology for maintaining inherent data security and effective privacy. This is creating another concern as expansion of application of this technology is inviting more and more novel security threats targeted on the blockchain [31]. Hence, it is now time for researchers to focus on strengthening the security of blockchain. For the same reason, a threat model has been presented in Section 3.3.1.

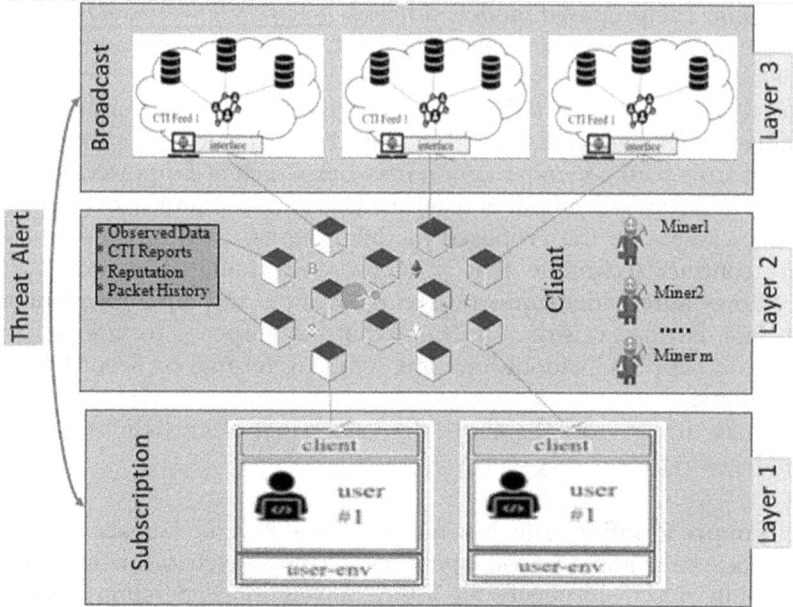

FIGURE 3.3
Threat model for blockchain system [32].

3.3.1 Threat Model for Blockchain

The generic architecture of a blockchain-based cyber threat intelligence system consists of a three layer process, as shown in Figure 3.3 [32].

1. **User Layer:** This layer consists of consumers and contributors as users. The layer basically intends to collect data from the user's environment, such as firewalls and IPS. This layer enables the sharing of threat-related data which is observed with reference to standard specifications. Additionally, the user can use a data parser in order to collect and pre-process the data pertaining to the user's environment, which can also be considered as an extension of the environment's security system.

2. **Blockchain Network Layer:** The proposed framework employs blockchain for efficiently managing and sharing data. This layer consists of storage nodes and miner nodes. When security related

data is fetched or written to the user layer, information is also sent to the blockchain network. Miner nodes get rewarded by checking user's requests. Further, smart contracts are used to process any data pertaining to users that need to fetch data from feeds, thus, ensuring the traceability and integrity of the data.

3. **Feed Layer**: This layer provides information related to cyber threat intelligence to consumers in the list. Here, each feed has its data evaluation function that collects data reported to the blockchain network, in addition to determining the validity of the collected data. Further, the feed may also generate a warning regarding data contributor if it is obtained from a malicious party, thus, lowering confidence in the malicious contributor. Consequently, the evaluation function of feeds decreases, which makes the contribution of malicious user nearly impossible.

Various applications and related cyber security attacks which can be handled by blockchain are demonstrated in Figure 3.4.

Now, after discussing the threat models for ML and cyber security, authors present a comparative study in Table 3.3 to enhance the understanding of the readers.

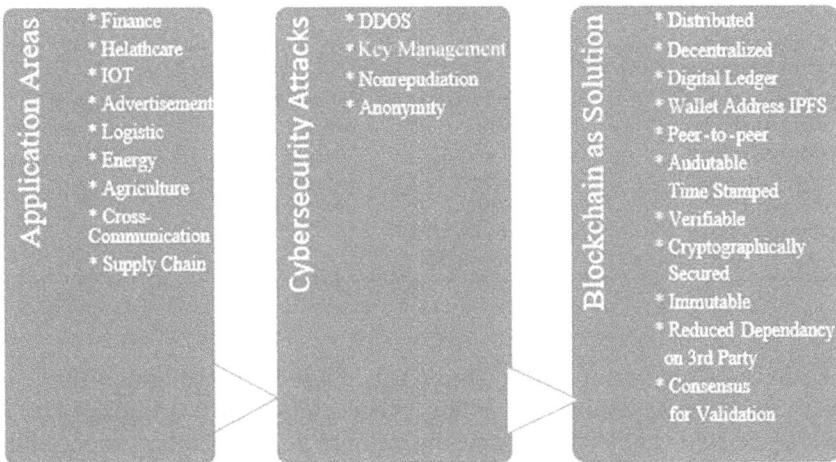

FIGURE 3.4
Application areas, cyber security attacks and blockchain characteristics as solutions.

TABLE 3.3

Comparative Study

Title	Findings	Limitations	Mitigation
Blockchain-based DDoS mitigation using ML techniques [33]	Identify if the incoming packet is malicious or not using ML and store the blacklisted IP address using blockchain		Mitigate DDoS attacks using blockchain and ML
ML adoption in blockchain-based smart applications: The challenges, and a way forward [34]	Use ML and blockchain in smart applications	Infrastructure availability, quantum resilience and privacy issues	
A blockchain solution for enhancing cyber security defense of IoT [35]	The decentralized structure to support cyber security defense mechanisms in smart cities of system	High cost	To reduce the cost of gas value, can use private blockchain network
A use case in cyber security based in blockchain to deal with the security and privacy of citizens' and smart cities' cyberinfrastructures [36]	Smart city with IoT used to overcome failures and cyberinfrastructure hacker's attacks	Scalability and social acceptance	Scalability can improve with lightened network, need to regulate social laws
Application of rank-weight methods to blockchain cyber security vulnerability assessment framework [37]	NIST cyber security framework used enhanced cyber security vulnerabilities using various rank-weight methods	Not in real time use, subjective analysis required at initial stages	Can integrate various methods into proposed framework

From the above comparative study, it is obvious that although there are plenty of approaches for preventing cyber-attacks, each approach has its own associated challenges. Hence, the authors in this chapter propose integration of ML and blockchain for mitigating the network attacks as discussed in the subsequent section.

3.4 Proposed Approach

Authors in this chapter propose a hybrid approach that integrates ML and blockchain in order to mitigate cyber security threats. An attacker always aims to unveil network vulnerabilities in order to initiate an attack, whereas researchers focus on devising new solutions and approaches to prevent attacks. Now, from the above discussion, it is evident that emerging technologies like ML and blockchain open new avenues for enhancing cyber security and protecting data privacy from various attacks.

In order to establish the effectiveness of this hybrid approach, the authors present work by various researchers. The case study presented in [38] discusses the data preserving AI learning model using blockchain technology for cyber security in an open network. The authors in [38] also consider some cases for inaccuracy of AI learning data in terms of cyber security. The authors also discuss the need to learn data management ahead of applying ML concepts, through analysis of cyber security attacks techniques, in order to prevent cyber-attacks and data degradation. The framework proposed in [38] uses a blockchain-based ML environment model to verify the integrity of learning data.

The network may experience various attacks like Poisoning attacks, Impersonation attacks, Evasion attacks, or Inversion attacks. In order to mitigate these attacks, blockchain may be employed for data verification ahead of ML steps, as illustrated in Figure 3.5. The proposed model works on data provided by actors' and, further, uses ML models and blockchain. The proposed method ensures that the raw data supplied by providers should not be tampered with during processing. The proposed approach uses blockchain to handle raw data and processes it ahead of storing on IPFS. The data is verified using a log of verified data which is stored on blockchain network. Further, verified data is forwarded to the ML model for data preprocessing and finally preprocessed data is verified by the verification server. The hashed values and related data are maintained on blockchain and IPFS respectively. in order to avoid various attacks.

Hence, the combination of ML (AI mode) and blockchain is fine tuned in the proposed approach for data preservation and verification purposes, in order to minimize the cyber-attacks and maximize the network security. The decentralized nature of the proposed model prevents single point failure, thus, maximizing the availability of the model. The proposed model achieves the benefits of both approaches (ML and blockchain) and thus enhances the efficiency of the proposed model.

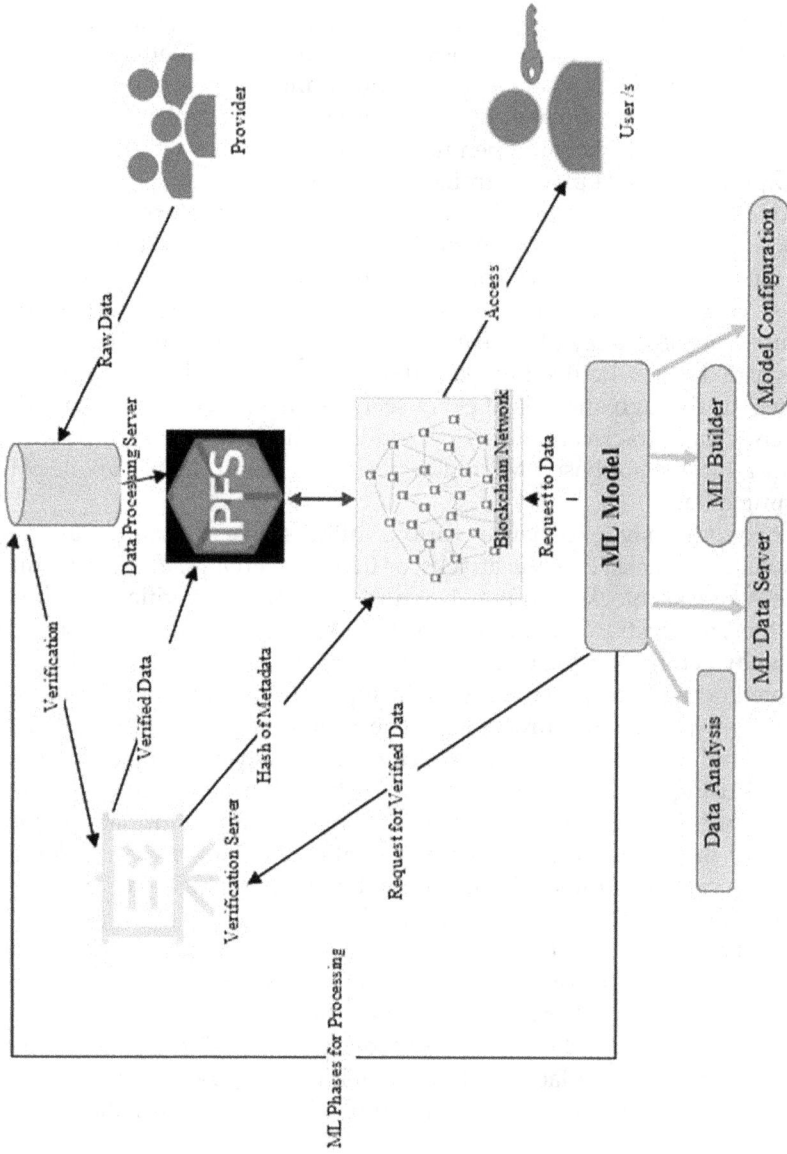

FIGURE 3.5
ML and verification model based on blockchain.

3.5 Conclusion and Future Work

The authors in this chapter have studied the applicability of ML and block-chain in preventing network vulnerabilities. While machine learning has experienced applications in the field of cyber security during the past few decades, blockchain has also been widely accepted as a promising technology in this regard. The competence of blockchain in cyber security is obtained as a result of decentralized structures and the consensus algorithm. Although blockchain has been observed to yield promising performance for cyber security, recently attackers have been targeting the blockchain so as to damage its structure. Hence, a further challenge facing researchers is to ensure the security of blockchain and, consequently, numerous researchers have been working in the direction of maintaining security at different layers of blockchain.

References

[1] Sajal S., Jahan I., Nygard K. A Survey on Cyber security Threats and Challenges in Modern Society. *IEEE International Conference on Electro Information Technology (EIT)* (pp. 525–528). doi:10.1109/EIT.2019.8833829

[2] Passeri, P. *Cyber Attacks Statistics.* Paolo Passeri, 13 April 2021. http://www.hackmageddon.com/category/security/cyber-attacks-statistics/. Accessed 28 April 2021.

[3] Saravanan A., Sathya Bama S. A Review on Cyber Security and the Fifth Generation Cyberattacks. *Oriental Journal of Computer Science and Technology* 2019, 12(2), 50–56. doi:10.13005/ojcst12.02.04

[4] Truong T., Diep Q., Zelinka I. Artificial Intelligence in the Cyber Domain: Offense and Defense. *MDPI Journal of Symmetry* 2020, 12, 410. doi:10.3390/sym12030410.

[5] Sagar R., Jhaveri R., Borrego C. Applications in Security and Evasions in Machine Learning: A Survey. *MDPI Journal of Electronics* 2020, 9, 97. doi:10.3390/electronics9010097

[6] Sarker I., Kayes A. S. M., Badsha S., et al. Cybersecurity Data science: An Overview from Machine Learning Perspective. *Journal of Big Data* 2020 7, 41. doi:10.1186/s40537-020-00318-5

[7] Truong T., Diep Q., Zelinka I. Artificial Intelligence in the Cyber Domain: Offense and Defense. *MDPI Journal of Symmetry* 2020, 12, 410. doi:10.3390/sym12030410.

[8] SaiSindhuTheja R., Shyam G. A Machine Learning Based Attack Detection and Mitigation Using a Secure SaaS Framework. *Journal of King Saud University – Computer and Information Sciences* 2022, 34(7), 4047–4061. doi:10.1016/j.jksuci.2020.10.005

[9] Bagaa M., Taleb T., Bernabe J., Skarmeta A. A Machine Learning Security Framework for IoT Systems. *IEEE Access* 2016, 4, 1–12. doi:10.1109/ACCESS.2020.2996214

[10] Shaukat K. et al. A Survey on Machine Learning Techniques for Cyber Security in the Last Decade. *IEEE Access* 2020, 8, 222310–222354. doi:10.1109/ACCESS.2020.3041951

[11] Chowdhury A. Recent Cyber Security Attacks and Their Mitigation Approaches – An Overview. *CCIS* 2016, 651, 54–65. doi:10.1007/978-981-10-2741-3_5

[12] Zewdie T., Girma A. IOT Security and the Role of AI/ML to Combat Emerging Cyber Threats in Cloud Computing Environment. *Information Systems* 2020, 21(4), 253–263.

[13] Mohasseb A., Aziz B., et.al. Predicting CyberSecurity Incidents Using Machine Learning Algorithms: A Case Study of Korean SMEs. *Proceedings of the 5th International Conference on Information Systems Security and Privacy.* 2019. doi:10.5220/0007309302300237

[14] Dasgupta D., Akhta Z., Sen S. Machine Learning in Cybersecurity: A Comprehensive Survey. *Journal of Defense Modeling and Simulation: Applications, Methodology, Technology.* 2022. 19, 1, 57–106. doi:10.1177/1548512920951275

[15] Rege M., Mbah R. Machine Learning for Cyber Defense and Attack. *Data Analytics 2018: The Seventh International Conference on Data Analytics.* 2018 (pp. 73–78).

[16] Yaga D. (NIST), Mell P. (NIST), Roby N. (G2), Scarfone K. (Scarfone Cybersecurity). NISTIR 8202 Blockchain Technology Overview. October 2018. https://csrc.nist.gov/publications/detail/nistir/8202/final

[17] Technical Committee: ISO/TC 307 Blockchain and Distributed Ledger Technologies. ISO 22739:2020 Blockchain and Distributed Ledger Technologies— Vocabulary. July 2020. https://www.iso.org/standard/73771.html

[18] About R3 Blockchain/DLT 101. *r3.com(Blockchain101).* https://www.r3.com/blockchain-101/

[19] Nakamoto, S. Bitcoin: A Peer-to-Peer Electronic Cash System, 2008. [Online]. Available: https://bitcoin.org/bitcoin.pdf.

[20] Taylor P. J. A Systematic Literature Review of Blockchain Cyber Security. *Digital Communications and Networks* 2020, 10.

[21] Julien Legrand, Security Governance Lead at Neat. The Future Use Cases of Blockchain for Cybersecurity. *Cyber Management Alliance*, September 04, 2020. https://www.cm-alliance.com/cybersecurity-blog/the-future-use-cases-of-blockchain-for-cybersecurity

[22] S. Bansod, L. Ragha. Blockchain Technology: Applications and Research Challenges. *2020 International Conference for Emerging Technology (INCET)*, Belgaum, India, 2020, pp. 1–6, doi: 10.1109/INCET49848.2020.9154065.

[23] Wani S., Imthiyas M., Almohamedh H., Alhamed K. M., Almotairi S., Gulzar Y. Distributed Denial of Service (DDoS) Mitigation Using Blockchain—A Comprehensive Insight. *Symmetry* 2021, 13(2), 227. doi:10.3390/sym13020227

[24] Singh R., Tanwar S., Sharma T. Utilization of Blockchain for Mitigating the Distributed Denial of Service Attacks. *Security Privacy* 2019, 3, e96. wileyonlinelibrary.com/journal/spy2. doi:10.1002/spy2.96

[25] Beuria M. K., Sharma P. K., et al. Applicability of Blockchain towards Mitigation of Distributed Denial of Service Attack in IoT. *European Journal of Molecular & Clinical Medicine* 2020, 07(10), 3432–3442. ISSN 2515-8260

[26] Lage O., Diego S., et al. *Blockchain Applications in Cybersecurity*. IntechOpen Limited, London UK, November 2019. doi:10.5772/intechopen.90061

[27] ElMamy S., Mrabet H., et al. A Survey on the Usage of Blockchain Technology for Cyber-Threats in the Context of Industry 4.0. *Sustainability* 2020, 12, 9179. doi:10.3390/su12219179

[28] Hajizadeh M., Afraz N. Collaborative Cyber Attack Defense in SDN Networks Using Blockchain Technology. *IEEE Conference on Network Softwarization (NetSoft)*. 2020. doi:10.1109/NetSoft48620.2020.9165396

[29] Graf R., King R. Neural Network and Blockchain Based Technique for Cyber Threat Intelligence and Situational Awareness. *2018 10th International Conference on Cyber Conflict (CyCon)* (pp. 409–426). IEEE, 2018, May.

[30] Dai F., Shi Y., Meng N., Wei L., Ye Z. From Bitcoin to Cybersecurity: A Comparative Study of Blockchain Application and Security Issues. *2017 4th International Conference on Systems and Informatics (ICSAI)* (pp. 975–979). IEEE, 2017, November.

[31] Wang H., Wang Y., Cao Z., Li Z., Xiong G. An Overview of Blockchain Security Analysis. *China Cyber Security Annual Conference* (pp. 55–72). Springer, Singapore, 2018, August.

[32] Gong S., Lee C. BLOCIS: Blockchain-Based Cyber Threat Intelligence Sharing Framework for Sybil-Resistance. *Electronics* 2020, 9, 521. doi:10.3390/electronics903052

[33] Manikumar D. V. V. S., Uma M. B. *2020 Second International Conference on Inventive Research in Computing Application*. doi:10.1109/ICIRCA48905.2020.9183092

[34] Tanwar S., Bhatia Q., et al. Machine Learning Adoption in Blockchain-Based Smart Applications: The Challenges, and a Way Forward. *IEEE Access* 2016. doi:10.1109/ACCESS.2019.2961372

[35] Giannoutakis K. M., et al. A Blockchain Solution for Enhancing Cybersecurity Defence of IoT. 2020 *IEEE International Conference on Blockchain (Blockchain)* 2020, 490–495. doi:10.1109/Blockchain50366.2020.00071

[36] Mora O. B., Rivera R., Larios V. M., Beltrán-Ramírez J. R., Maciel R., Ochoa A. A Use Case in Cybersecurity Based in Blockchain to Deal with the Security and Privacy of Citizens and Smart Cities Cyberinfrastructures. *2018 IEEE International Smart Cities Conference (ISC2)* (pp. 1–4). 2018. doi:10.1109/ISC2.2018.8656694

[37] Gupta Gourisetti N., Mylrea M., Patangia H. Application of Rank-Weight Methods to Blockchain Cybersecurity Vulnerability Assessment Framework. *2019 IEEE 9th Annual Computing and Communication Workshop and Conference (CCWC)* (pp. 0206–0213). 2019. doi:10.1109/CCWC.2019.8666518

[38] Kim, J., Park, N. Blockchain-Based Data-Preserving AI Learning Environment Model for AI Cybersecurity Systems in IoT Service Environments. *Applied Sciences* 2020, 10, 4718. doi:10.3390/app10144718

Section II

Defending Against Cyber Attack Using Machine Learning

4

Detection of Spear Phishing Using Natural Language Processing

Deep Gandhi, Jash Mehta and Ramchandra Mangrulkar
University of Mumbai, Mumbai, India

CONTENTS

4.1 Introduction

Spear phishing is one of the most common forms of attack and it targets a particular user or a group of users through malicious emails. Phishing can be defined as a pernicious form of cyber-attack, conducted to persuade the targeted user to perform actions for the benefit of the attacker, with the help of socially engineered messages through digital communication channels [1].

The attackers tend to target the weakest link in order to breach highly secure systems successfully. Often, the weakest link in such systems is the users [2]. Frequently the attack exploits individual characteristics, which are specific to individuals or their organizations, with the motive of establishing trust [3]. According to [4], a user is 4.5 times more likely to fall for a phished email which is sent by an attacker pretending to be a trustworthy source such as an existing contact, boss or friend, than to fall for ther kinds of standard phishing attempts.

DOI: 10.1201/9781003408307-6

Due to the Covid 19 pandemic, a majority of work across organizations has shifted online and all the daily workplace communications have also been moved to email systems. This has given rise to an open exploitation channel for attackers. Spear phishing has been on the rise for a long time and due to the increase in usage of emails, attackers gain access to many more opportunities than before to open a channel of data leakage by pretending to be some senior official and then getting the victims to reveal confidential information about the organization or, at times, about the victims themselves. This has become an open risk which is increasing day-by-day as major organizations have declared eternal work-from-home schedules.

As proposed by [5], the most vulnerable factor which makes a victim suitable for a spear-phishing attack is the conscientiousness personality trait. Since the Covid 19 pandemic led to the loss of many jobs, people are fearful being laid off and had a considerable impact on their minds: it is this fear which motivates every employee to be conscientious. Thus, when faced with such situations, employees do not think twice, end up revealing confidential information to the attacker.

Keeping these problems in mind, the authors propose a system to profile communication within workplace emails and then to develop a speaking style profile for every person within a single model, in order to detect any anomaly whenever someone tries to impersonate the aforementioned person. In order to mitigate phishing attacks, an approach similar to an email filtering system can be used [6]. A risk score for each email can be calculated, and the email can be discarded if the score surpasses a pre-specified threshold. The threshold value should be carefully selected since a low threshold value would lead to the discarding of non-malicious emails as well. Thus, finding an optimal balance between security and usability is of paramount importance [7].

4.2 Literature Review

The authors study how [5] deals with the psychological factors that are a key factor in information leakage. Conscientiousness personality trait has shown the most correlation with the likelihood of a user falling prey to a spear phishing attack. Since the vulnerability to phishing depends on the personality and not on the awareness of the user, it is difficult to predict phishing in advance. Security effectiveness can be increased if a targeted approach is taken.

Along with this, [8] also discusses the reason behind the lower efficacy and the broken feedback mechanisms of the spear-phishing systems. This is because of low reporting of spear-phishing emails by users, due to factors such as self-efficacy and the fear of wrongly reporting a spear-phishing email. This makes it difficult to learn more about the science of spear phishing and security in general.

The authors also explore how in [9], the authors experiment with a personalized opening line in a spear-phishing email and detect that 29% of those provide personally identifiable information in a spear-phishing attack. The findings of the given paper also explored how age affected whether employees revealed information and found age to be an insignificant factor. However, it was found that a personalized opening to the message was a major factor which led to employees revealing identifiable sensitive information.

This provides conclusive proof that an automated system is required to deal with spear-phishing attacks, along with spreading awareness among users. The authors studied the system proposed by [10] which scrutinize the email content in order to detect stylometric features and then, to detect anomalies in the email. This paper also dealt with how probabilistic models were able to distinguish phishing attempts based on stylometric features in a closed employee system of 20 users. Thus, building on such automated approaches, a highly scalable approach is proposed in this chapter, as a system of only 20 users is not feasible in a high-end organization with multiple departments.

4.3 Dataset

For the task of identification of spear-phishing emails, a subset of the Enron email dataset [11] is used in the proposed approach. The Enron corpus is a large set of email messages which was released publicly by the Federal Energy Regulatory Commission during its investigation concerning the Enron corporation.

This corpus contains the organized data of about 150 employees, most of whom were from the senior management at Enron. A total of around 500,000 messages are present in the corpus which makes it the only large dataset of real-world emails which is open-sourced for public use, according to the best of the authors' knowledge. Each folder in the dataset provides all the details about the email such as the information of the sender and the receiver, date and time of the email, contents of the email and subject among others. The dataset consists of official emails from personnel at all levels of the corporation, making it perfect for the task at hand.

A subset of the Enron dataset having 72,932 emails is used in the proposed system.

4.4 Data Preprocessing

The Enron email dataset contains data that is organized into folders – consisting of files along with the corresponding messages. Each of these messages contains all the technical information about an email such as its mime version, content type, the email addresses of sender and receiver, the subject of the email, the main content, and so forth. Since all of this information is contained inside one single message, it is separated and segregated into relevant columns.

The columns which had information such as the type of content, content transfer encoding, mime version, and so on- were dropped because they had very few values. Moreover, information other than date and time, sender's and receiver's email addresses, subject of the email, and its content, was discarded since it is irrelevant to the proposed framework for the task of identifying spear-phishing emails. The date in all the emails is converted to a date-time format for efficient attribute extraction which would help in further preprocessing.

The subset of the Enron email dataset which is used in this task contains almost 73,000 emails. Upon thorough exploratory analysis of this data, some interesting conclusions can be drawn. As shown in Figure 4.1, the dataset contains emails exchanged in the early 2000s. Most of the emails were sent on weekdays and during working hours as expected.

Figure 4.2 shows the pair-wise relationships in the dataset between the number of emails sent by a user, the word count of the subject and the word count of the content. A social network analysis of email sender and recipients is also conducted to determine the frequency with which two particular individuals exchange emails. The findings of this analysis can be seen in Figure 4.3. There are a few users who send a lot of emails to themselves. Thus, it is quite interesting to look at the differences between the emails that users send to themselves as compared to those that are sent to others.

To process the main content present in the emails, a few preprocessing steps are carried out, before feeding it into the proposed model. For the whole process, a smaller batch size is used since the dataset is very large and some of the operations which are performed are computationally costly like training the language model which uses a lot of GPU. There are two main preprocessing steps: tokenization and numericalization. This is accomplished with the help of fastai's [12] Data Block API.

Tokenization: This is the breaking down or splitting of raw sentences into words or more specifically, tokens. Splitting the string into spaces is the simplest way of tokenizing a sentence, however, it is avoided in this task because a lot of useful information is lost while doing so. Instead, the punctuation, contrast words, and other minute details are taken care of during tokenization. Numerous special tokens are also added during tokenization. Words

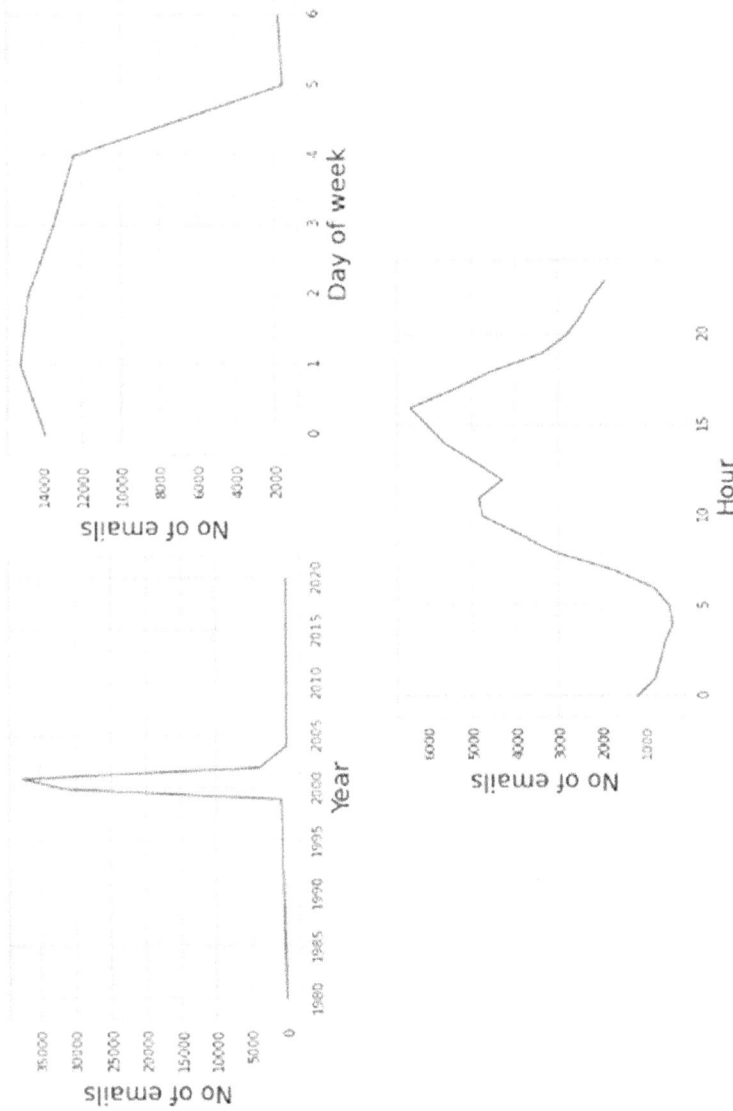

FIGURE 4.1
Pairwise relationships of emails.

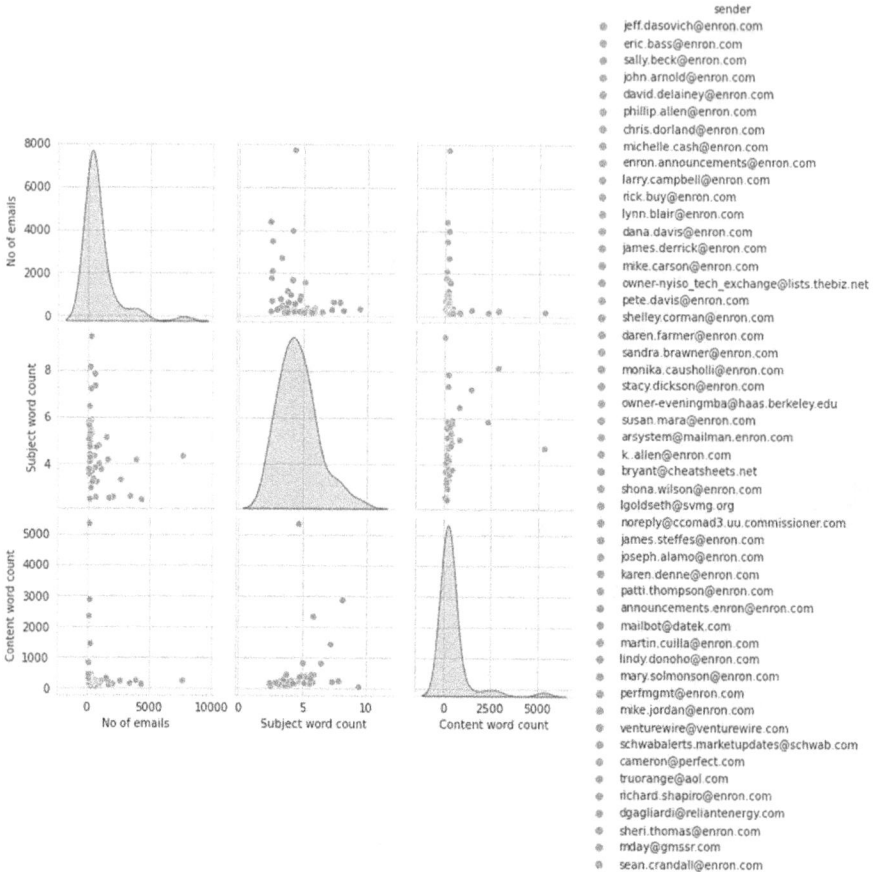

FIGURE 4.2
Distribution of emails.

which are very rare and are not a part of the model's vocabulary, are assigned a special token UNK. Another special token PAD is used for padding to deal with texts of varied length while regrouping them in a batch. If a particular word is written in all capital letters, it is preceded by TK_UP. This helps in preserving information about semantics and indicates that the word was emphasized. Several other tokens are used as well.

Numericalization: This is relatively easier since it involves assigning a unique id to each token obtained after tokenization and mapping each of the obtained tokens to the corresponding ids. In simpler terms, it is equivalent to assigning an integer value to each token.

After the preprocessing is done, 20% of the entire data is put into a validation set, after the data has been randomly shuffled. Shuffling is done to

sender	recipient1	count
pete.davis@enron.com	pete.davis@enron.com	627
eric.bass@enron.com	shanna.husser@enron.com	491
sally.beck@enron.com	patti.thompson@enron.com	388
eric.bass@enron.com	jason.bass2@compaq.com	374
jeff.dasovich@enron.com	nancy.sellers@robertmondavi.com	362
enron.announcements@enron.com	all.houston@enron.com	360
enron.announcements@enron.com	all.worldwide@enron.com	334
owner-eveningmba@haas.berkeley.edu	eveningmba@haas.berkeley.edu	301
jeff.dasovich@enron.com	joseph.alamo@enron.com	297
david.delainey@enron.com	kay.chapman@enron.com	290

FIGURE 4.3
Sender and receiver analysis.

ensure that the validation set contains emails from all users and not only from a few of them. A predetermined random seed is used so that the splitting remains deterministic at any given time. The remaining 80% of the data is the training set.

4.5 Textual Anomaly Detection System

The proposed system for the chapter includes two modules to identify phishing attempts in a given email. These two modules are Style Detection and Sender Detection. Each of these modules is further explained in detail.

4.5.1 Style Detection

The authors implement a ULMFiT approach [13] to identify the real sender of the email. The ULMFiT approach can be observed in Figure 4.4.

According to this approach, to establish some syntactic and semantic structures for the target language (English, in this case), a language model is trained on the WikiText-103 corpus [14]. This language model is trained on the AWD-LSTM model [15] which is further explained in detail. After initial training, this language model is further fine-tuned on the given email corpus of the Enron dataset. All the email content is taken and fine-tuned on the

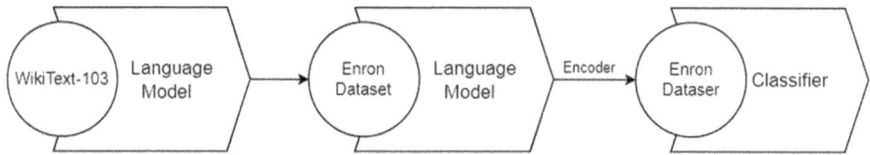

FIGURE 4.4
ULMFiT-based email identification approach.

pretrained model. This helps in increasing the vocabulary and also helps the language model learn some nuances of email communication which may not be found on the WikiText-103 dataset [14].

After this fine-tuning is complete, the encoder from the language model is taken and then used to build a classifier that would be further responsible for the writing style identification of various employees in the company. This is done as the pre-training on both the corpora allows the encoder to then train on the writing style of individual authors, rather than still training and building upon language structures during the training process of the classifier. Thus, as discussed earlier, the Enron dataset used for this chapter consists of emails from 4591 separate employees at various stages and thus, this helps in also determining the scalability of the system while identifying individual authors. Since the encoder of the AWD-LSTM is used, it is considered a wise choice to use the same architecture for the classifier. The classifier is further trained and thus, it learns to classify the individual employees based on the text provided. The intricate details regarding the architecture and the hyper parameters are discussed further.

4.5.1.1 Architecture of AWD-LSTM

The AWD-LSTM as mentioned in [15] is an enhancement of the present LSTM architecture by the usage of DropConnect as proposed by [16]. According to the [15], the optimizer update by the SGD proposed does not return the latest updated weights. As proposed, an average of the weights of the previous iterations is returned. This method is known as ASGD in the paper and helps in dealing with a possible overfitting issue in the model. The DropConnect proposed in the [15] means that instead of dropping a random subset of activations as proposed by [17], the connections are dropped to prevent overfitting. This is because the loops of LSTMs stacked together may result in the data overfitting. DropConnect may seem like Dropout, however, the significant distinction between the two is that, as proposed in [17], Dropout chooses a few units arbitrarily utilizing a given likelihood and sets their yields to zero, and DropConnect works by setting singular loads to zero aimlessly rather than the total node. This aids in protecting the quantity of hubs and, furthermore, presents the arbitrariness for 'co-transformation' that Dropout focuses on. The DropConnect technique, hence, empowers the LSTMs to get contributions from various subsets of the past layer. As mentioned in [15], the AWD LSTM achieved a great result on the Penn Treebank and WikiText-2 with perplexity scores of 57.3 and 65.8 respectively. Due to this enhanced performance by the model, it is found to be suitable for usage in both the language model as well as the classifier for writing style identification of emails in a specific organization.

4.5.1.2 Training of the Proposed Model

The training process begins with the initial training of the language model and then fine-tuning the said language model on the email dataset. For ease of computation, the AWD-LSTM language model to be used as a pre-trained model on WikiText-103 is taken from [13]. This pretrained language model is then fine-tuned with the Enron dataset that has been cleaned and preprocessed. During this fine-tuning processing, hyperparameter tuning is carried out to get optimum performance of the language model. The hyperparameters are as follows:

1. *Discriminative Learning Rates*: This was proposed in [13]. According to this methodology, different learning rates are used at different stages of a model to be fine-tuned during the transfer learning process. This is useful as it helps in maintaining the pretrained weights while training the model on new data. Thus, in this project, the language model fine-tuning is done using the learning rate value of 0.02 for the newer layers and the learning rate value of 0.0067 for the older values of the model. The performance of learning rate against the loss for the language model can be seen in Figure 4.5.

2. *One-Cycle Scheduling*: Along with this discriminative learning rate, it has been ensured that the optimum performance is achieved while training. This can be completed using the one-cycle scheduling approach as proposed in [18]. According to [18], weight decay is used for regularization. Along with this, the momentum as proposed in [18] is set as (0.8, 0.7) arbitrarily as this is found to be the optimum value for the training procedure.

3. *Mixed-Precision Training*: To diminish computational complexity and repetition, the mixed-precision approach is proposed in [19]. By and large, all the parameters are stored as 32-bit floats. In any case, one can utilize half-precision to diminish these floating points to 16-bit floats as this change is relied upon to not react excessively to a distinction in the weights. It empowers the model to utilize twofold the batch size and train more quickly. In any case, because of this reduction, issues emerge like vanishing gradients, loss overflow and

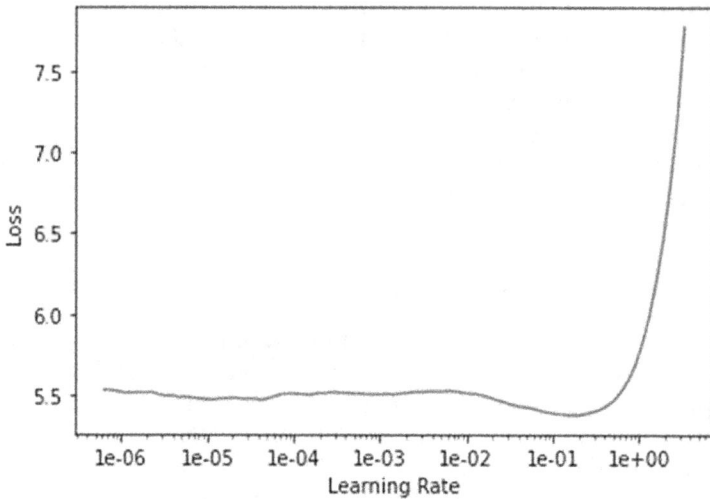

FIGURE 4.5
Learning rate vs loss graph for language model.

so on. Along these lines, a mixed-precision approach is utilized, as referenced in [19]. This prompts a few elements being 32-bit floats and some as 16-bit floating points. Thus, the important information such as updating the weights is carried out with 32-bit floats and some other tasks such as a forward pass to be carried out with half-precision. Along with this, mixed-precision also prevents profligate training procedures by keeping half of the bits that aren't required on the Central Processing Unit (CPU) and the other half on GPU. This aids in protecting Graphics Processing Unit (GPU) memory and furthermore lessens GPU costs. Subsequently, this prompts quicker, yet just as economical, training.

After using all of these tricks to improve the language model training, the fine-tuned encoder of this language model is taken out and then used as the encoder during the initialization of the classifier. This classifier is the one that would predict the author based on the writing style of the email. During the training procedure, similar hyperparameters are also used for the classifier as the same architecture of the model is to be used. However, during this classifier training, the learning rate used is 0.01 for the newer layers and 0.0033 for the older pretrained layers. This helps, as discussed earlier. The performance of the learning rate against the loss for the training of the classifier is given in Figure 4.6.

Thus, the training process of the language model, as well as the classifier, is carried out using the most optimum training practices.

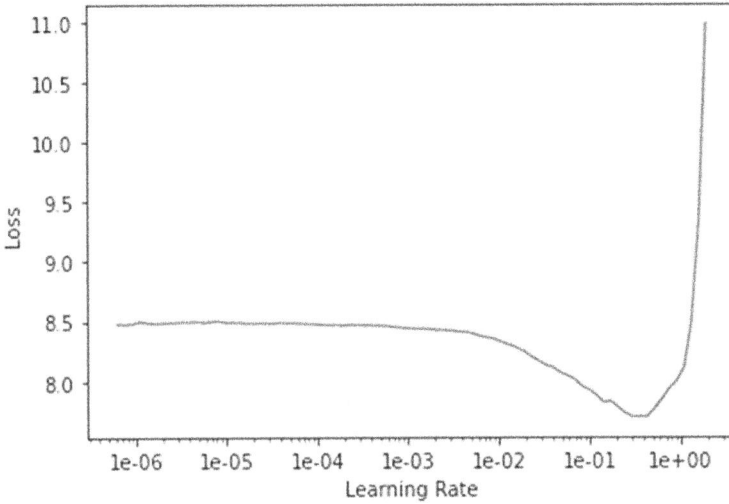

FIGURE 4.6
Learning rate vs loss graph for classifier.

4.6 Experimentation and Results

Table 4.1 demonstrates the results achieved while performing the given task. It must be noted that the accuracy of the given style identification is a mere 53.36%. This is because of two reasons which can be resolved completely, depending on the organization. The first reason is that the Enron subset used for training consists of 4,591 individual employees. The sheer size of the number of employees in the dataset affects the accuracy. This can be resolved for smaller organizations as they would have a significantly smaller number of employees.

However, for a bigger organization, using additional computational resources to keep training the classifier is recommended. This experiment was limited due to the restricted availability of GPU resources. Thus, it would

TABLE 4.1

Results and Description

Model	Description	Accuracy (%)
ULMFiT-based Language Model	Fine-tuned on all of the text from emails after being trained on the WikiText-102 corpus.	51.1
ULMFiT-based Classifier	Trained on the labels from the dataset using the encoder of the LM	53.36

be resolved if bigger organizations keep training the classifier continuously for a higher number of epochs.

It can be observed based on the results that the language model is trained well and achieves an accuracy rate of 51.1%. This is also because of the large corpus of emails of the dataset. The variation in writing styles does exist, however, the vocabulary of the emails for the dataset does not differ much and thus, would lead to the language model grasping the syntactic and semantic dependencies of the language better and thus, providing better accuracy.

It must also be noted that the low performance of the classifier might be due to the large text length of each email. This might lead to some recurrence problems in the network. Thus, it would also benefit if a dataset of short emails is available in any given organization between the same pair of users. This would help the classifier in capturing the writing style better and thus, increase the accuracy significantly.

As demonstrated in [20], a higher accuracy such as 94% is attained on the Enron dataset while considering only ten unique authors. However, the problem with this is that such systems would not be useful in the real world as they are not scalable for a high number of authors (here 4,591) which is observed in [20]: when the number of authors is increased from 10 to 25, the performance degrades directly to an 81% accuracy level. This would result in large-scale discrepancies between the research results and the practical work results in a real-world environment. This is a major reason why the number of authors was not truncated in this project.

4.7 Future Applications

As scalability was discussed in detail in the experimentation section, it would be great to further advance this system such that scalability is not an issue at all. This would be useful if the current infrastructure is further deployed and trained using federated learning as demonstrated in [21]. Federated learning would, thus, utilize the distributed computing approach and help in training the model on the complete dataset without hesitation on an increased number of unique employees. This would also help in shared learning, based on localized regions, and thus, creating geographical clusters within the federated learning ecosystem would further help in identifying the employee and thus, detecting the phishing attempt.

The federated learning approach would also protect private and sensitive contents of the mail from being shared on the collaborative central server. This is the major issue dealt with in [21]. Since the emails of any organization consist of sensitive contents and sharing between departments is also restricted, it would be beneficial if the privacy of these emails is protected. However, since the system requires the content to train and identify the real author, the future version of the system could incorporate federated learning to deal with the issue as the localized training of such a system would result in the email never leaving the system of the user.

4.8 Conclusion

In summary, the chapter deals with how the problem of spear phishing is handled by using a ULMFiT-based Natural Language Processing (NLP) system. Even though the problem has amplified during the Covid-19 lockdown phase, a resourceful solution has been found to deal with the problem, and thus, limit such phishing attempts, save the user data and protect the integrity of the organization. Thus, the synergy of NLP and cyber security also proves the co-dependency of both the fields in deal with crucial problems in society.

References

[1] Khonji, Mahmoud, Youssef Iraqi, and Andrew Jones. "Phishing detection: A literature survey." *IEEE Communications Surveys & Tutorials* 15, no. 4 (2013): 2091–2121.

[2] Sasse, Martina Angela, Sacha Brostoff, and Dirk Weirich. "Transforming the 'weakest link'—A human/computer interaction approach to usable and effective security." *BT Technology Journal* 19, no. 3 (2001): 122–131.

[3] Hong, Jason. "The state of phishing attacks." *Communications of the ACM* 55, no. 1 (2012): 74–81.

[4] Jagatic, Tom N., Nathaniel A. Johnson, Markus Jakobsson, and Filippo Menczer. "Social phishing." *Communications of the ACM* 50, no. 10 (2007): 94–100.

[5] Halevi, Tzipora, Nasir Memon, and Oded Nov. "Spear-phishing in the wild: A real-world study of personality, phishing self-efficacy and vulnerability to spear-phishing attacks." *Phishing Self-Efficacy and Vulnerability to Spear-Phishing Attacks* (January 2, 2015). Available at: SRN: https://ssrn.com/abstract=2544742 or http://dx.doi.org/10.2139/ssrn.2544742

[6] Laszka, Aron, Yevgeniy Vorobeychik, and Xenofon Koutsoukos. "Optimal personalized filtering against spear-phishing attacks." In *Proceedings of the AAAI Conference on Artificial Intelligence* 29, no. 1 (2015). doi:10.5555/2887007.2887140

[7] Sheng, Steve, Ponnurangam Kumaraguru, Alessandro Acquisti, Lorrie Cranor, and Jason Hong. "Improving phishing countermeasures: An analysis of expert interviews." In *2009 eCrime Researchers Summit*, pp. 1–15. IEEE, 2009.

[8] Kwak, Youngsun, Seyoung Lee, Amanda Damiano, and Arun Vishwanath. "Why do users not report spear phishing emails?" *Telematics and Informatics* 48 (2020): 101343.

[9] Bullee, Jan-Willem, Lorena Montoya, Marianne Junger, and Pieter Hartel. "Spear phishing in organizations explained." *Information & Computer Security* 25, no. 5 (2017): 593–613. https://doi.org/10.1108/ICS-03-2017-0009

[10] Duman, Sevtap, Kubra Kalkan-Cakmakci, Manuel Egele, William Robertson, and Engin Kirda. "Emailprofiler: Spearphishing filtering with header and stylometric features of emails." In *2016 IEEE 40th Annual Computer Software and Applications Conference (COMPSAC)*, vol. 1, pp. 408–416. IEEE, 2016.

[11] Klimt, Bryan, and Yiming Yang. "The Enron corpus: A new dataset for email classification research." In *European Conference on Machine Learning*, pp. 217–226. Springer, Berlin, Heidelberg, 2004.

[12] Howard, Jeremy, and Sylvain Gugger. "Fastai: A layered API for deep learning." *Information* 11, no. 2 (2020): 108.

[13] Howard, J., and Sebastian Ruder. "Universal language model fine-tuning for text classification." ACL (2018). Available at: https://aclanthology.org/P18-1031.pdf

[14] Merity, Stephen, Caiming Xiong, James Bradbury, and Richard Socher. "Pointer sentinel mixture models." *arXiv preprint arXiv:1609.07843* (2016).

[15] Merity, Stephen, Nitish Shirish Keskar, and Richard Socher. "Regularizing and optimizing LSTM language models." *arXiv preprint arXiv:1708.02182* (2017).

[16] Wan, Li, Matthew Zeiler, Sixin Zhang, Yann Le Cun, and Rob Fergus. "Regularization of neural networks using Dropconnect." In *International Conference on Machine Learning*, pp. 1058–1066. PMLR, 2013.

[17] Srivastava, Nitish, Geoffrey Hinton, Alex Krizhevsky, Ilya Sutskever, and Ruslan Salakhutdinov. "Dropout: A simple way to prevent neural networks from overfitting." *The Journal of Machine Learning Research* 15, no. 1 (2014): 1929–1958.

[18] Smith, Leslie N. "A disciplined approach to neural network hyper-parameters: Part 1—learning rate, batch size, momentum, and weight decay." *arXiv preprint arXiv:1803.09820* (2018).

[19] Micikevicius, Paulius, Sharan Narang, Jonah Alben, Gregory Diamos, Erich Elsen, David Garcia, Boris Ginsburg et al., "Mixed-precision training." *arXiv preprint arXiv:1710.03740* (2017).

[20] Nizamani, Sarwat, and Nasrullah Memon. "CEAI: CCM-based email authorship identification model." *Egyptian Informatics Journal* 14, no. 3 (2013): 239–249.

[21] Hard, Andrew, Kanishka Rao, Rajiv Mathews, Swaroop Ramaswamy, Françoise Beaufays, Sean Augenstein, Hubert Eichner, Chloé Kiddon, and Daniel Ramage. "Federated learning for mobile keyboard prediction." *arXiv preprint arXiv:1811.03604* (2018).

5

A Study of Recent Techniques to Detect Zero-Day Phishing Attacks

Sharvari Patil and Narendra M. Shekokar

University of Mumbai, Mumbai, India

CONTENTS

5.1 Introduction

Phishing is a fraudulent activity in which internet users are exposed to counterfeit websites or applications and their personal information is stolen [1]. During the COVID-19-triggered duration, the number of users using online banking operations increased, leading to a rise in phishing attacks. Around 60% of digital banking consumers are aged 18–26 years [2]. The customers in this age group are more likely to be conned by the phishers. Digital banking consumers also belong to the age group above 50 years, who are less aware of the various tricks used by the attackers to perform phishing attacks.

Various phishing detection approaches are the heuristic-based approach, blacklist approach, fuzzy-based approach, machine learning approach, CANTINA-based approach, and image-based approach [3]. Researchers have developed several techniques to protect users from phishing attacks, which are discussed in this chapter.

Threat actors are constantly looking for new ways to evade phishing detection systems. Attackers are now using artificial intelligence (AI) to build phishing URLs [4]. It has been observed that these AI-generated URLs are successful in bypassing the detection tools developed.

Researchers have used machine learning (ML), neural networks, AI, and deep learning (DL) methods to mitigate these types of attacks. We will study a technique developed by researchers that uses parallel execution of ensemble machine learning models with a multi-threading approach for the training and testing phases to detect AI-generated as well as human-made phishing URLs [5]. We will discuss these techniques in detail and then discuss a comparative analysis of the approach to identify the pros and cons of these techniques.

5.2 Phishing Detection Approaches

Phishing attacks can be identified using user education or automated detection tools. The user training method cannot be the sole contributor for detecting the attack. Since humans are bound to make errors or get trapped in the phishing web. Therefore, researchers have proposed various phishing detection methods to protect the user from being a victim of a phishing attack. Phishing detection strategies are categorized into education-based detection, list-based detection, heuristic-based detection, content-based detection, visual similarity-based detection and hybrid approaches.

5.2.1 Education-Based Detection

Users are educated or trained on different forms of phishing attacks. There are various ways in which this training can be provided to web users. These users can be trained using informative messages, messages from bank officials regarding possible frauds or certain training programs from cyber security bodies.

5.2.2 List-Based Detection

List-based detection is a static way of detecting phishing. This approach uses a list that includes fake/genuine or both types of URLs with labels.

The detecting mechanism then uses this list to alert the users. The blacklist maintains a list of fake websites in the database or on the cloud and the user is warned every time he/she tries to visit that website. The whitelist maintains the genuine URLs and prompts the users if they try to visit a URL other than the whitelisted list. This is an easy to implement technique and gives better throughput compared to the other methods. However, this technique fails when the attackers create a new website URL for attacking the users. These new URLs are not included in the database. This is an example of a zero-day phishing attack.

5.2.3 Heuristic-Based Detection

The heuristic-based approach considers a variety of phishing site features and calculates a threshold value to decide if the requested site is genuine or not [6]. Heuristic-based detection includes identifying features like grammar, length, domain age, spellcheck and scripting language. These features are further classified into two categories:

1. *Lexical Features*: The characteristics of the URL structure, like the length of the URL; number of dots(.); HTTPS protocol, presence of IP address; presence of special characters in the URL, or adding a prefix or suffix to the domain name.
2. *Third Party URL Features*: Third party URL features are obtained from external sources like age of the domain, Google page rank, popularity index.

These characteristics are used to create a rule-based scheme, or these characteristics are given weights, and a final threshold is determined to determine if the URL is genuine or not.

5.2.4 Content-Based Detection

In this detection technique, the matter of the web pages is analyzed, which involves the images and text material of the website [7]. This is a type of static approach where comparison is done based on the data from genuine websites.

1. *Image Analysis*
   Using image-processing algorithms, image analysis can be performed by comparing a website snapshot with the original webpage. Researchers have used logos, screenshots, and CAPTCHA for comparing with fake websites. This approach focuses on the visual content of web pages.

2. *Text Analysis*

Text analysis includes content in the web page, keywords, analysis of site logos and scripts. Text analysis is like static phishing detection where comparison is done between the fake website content and genuine web pages.

5.2.5 Hybrid Detection Technique

Hybrid detection is combination approach to detect zero-day phishing attacks. The techniques discussed are employed to retrieve the features of websites. These characteristics are fed into machine learning algorithms that determine whether a website is genuine or not.

5.3 Phishing Detection Using Machine Learning

Machine learning is a part of artificial intelligence (AI) that allows systems to learn and evolve without having to be specifically programmed [8]. Researchers use ML technology to detect zero-day phishing attacks. The general idea of this methodology is the development of a computer program that can access data and use it to learn prediction [9]. Machine learning algorithms are classified as supervised, unsupervised and reinforcement learning algorithms. We will discuss some approaches proposed by researchers using machine learning.

Sameen et al. have proposed a system that performs lexical evaluation for characteristic extraction. The system is deployed as the PhishHaven browser plugin. The authors have proposed novel techniques like URL Hit, considering HTML encoding as a feature for phishing detections. The system is divided into four components: URL Hit, Feature Extractor, Modelics and Decision Maker.

1. *URL Hit*: To extract confidential information from the user, attackers used tiny URLs. URL Hit approach is used to deal with these tiny URLs. This component redirects the URL requested by the user to the PhishHaven plugin where the tiny URL is converted to an actual URL. Feature extractor can then extract features from this actual URL.

2. *Feature Extractor*: This system does not use any third-party features of the URL. The system is based on lexical features of the URL and URL HTML encoding has been used as a lexical feature for detecting phishing URLs. The feature extraction process is conducted

in two stages, overall count and individual count, by dividing the URL into its components: segment, netloc, path, query and fragment.

3. *Modelics*: This component takes input from the feature extractor, but it works as a multithreaded model. Multiple threads are running simultaneously using the input from previous component. Each thread corresponds to a machine learning algorithm and gives output using that algorithm. Each thread sends their predicted result to the decision maker.

4. *Decision Maker*: This module uses the voting strategy which means the majority. It keeps track of the total number of expected results that are classified as phishing or genuine. Based on the total count, it then predicts the final label [5].

In [10] Sahingoz et al. used Natural Language Processing (NLP) features and seven classification algorithms to build a real-time anti-phishing system. The authors used a different feature set for each classification algorithm. The system classified the URL using NLP-based and word-based features. To assess the effectiveness of the proposed method, the researchers identified three types of feature sets: Word vectors, NLP-based features, and hybrid features. The dataset used for extracting these features is downloaded using script. In the data preprocessing stage, the URL is divided into its components and the words that are extracted are included in the word list. The objective of data preprocessing is to detect words like brand names, keywords and words which created using random characters. The Word Decomposer Module removes the numeric values from the words and performs a dictionary check of the words. If the words are found in the dictionary, they are put into the word list. The Random Word Detector Module deals with the random words that are added by the attacker. For the identification of random terms, the Markov Chain Model was used. In this process, the probabilities of letter pairs are calculated during the training stage. This calculated value is used as a new feature in the proposed system. The Maliciousness Analysis Module (MAM) uses Levenshtein Distance which measures the similarity between two strings by considering the deletions, insertions or substitutions operations in the strings. The output from this NLP module is used as a feature for identifying of counterfeit URLs. The researchers used seven different machine learning algorithms, including Decision Tree, Adaboost, K-Star, K-NN ($n = 3$), Random Forest, SMO, and Naive Bayes, as well as features like NLP-based features, word vectors, and hybrid features, to create a phishing detection system. The authors claim that using only NLP-based features in the Random Forest algorithm produces the best results.

5.4 Anti-Phishing Solutions Using Neural Networks/Deep Learning

Neural networks (NNs) are a sequence of algorithms structured like the neurons of a human brain. They are multilayers of neurons that we use to classify or cluster the datasets without labels [11]. In this section we will discuss various anti-phishing solutions using NNs or DL.

5.4.1 Solutions Using Neural Networks

Verma et al. proposed a system in which data is acquired from PhishTank websites and different types of attributes are gathered. Some attributes are collected by URL properties like length of URL, presence of special characters and others. Furthermore, this data is converted into categorical values, made in proper format and visualized. Then Artificial Neural Network (ANN) and Deep Belief Network (DBN)-based algorithms are implemented for classification of phishing URL. Algorithms used in this technique are back propagation and DBN. In back propagation technique error is calculated after every predicted output, and error correction formula is used to modify the weights. In DBN there are two steps: training in an unsupervised way and training in a supervised way. In the unsupervised way, probabilistic reconstruction of input is learnt by the DBN. This layer of unsupervised learning is called feature detectors on input. In training in the supervised way, DBN performs the classification [12].

The approach used by Adebowale et al. [13] is based on website phishing detection using the features of the site, content and their appearance. This paper proposes a web-phishing detection scheme based on an Adaptive Neuro-Fuzzy Inference System (ANFIS) that uses integrated features of text, image and frames. Image analysis is performed by the system using the Scale Invariant Feature Transform (SIFT) image-matching algorithm to detect phishing websites. The ANFIS model uses linguistic variables to identify phishing features. The study is deployed on a practical plug-in phishing toolbar implementation with suitable testing and validation. The Sugeno fuzzy model is used instead of Mamdani. The Sugeno fuzzy model membership functions are either linear or constant. The neuro-fuzzy inference system uses features as training and testing input data. To distinguish between legitimate, suspicious and phishing websites, the inference system generates fuzzy IF... THEN rules. The steps in the proposed system are as follows:

1. Website is the input to the systems, and features are extracted from these websites are text, images and frames.

2. Integrating the extracted features in websites to predict phishing activities, using a web browser plug-in.

3. The ANIFS is used for classification which determines if the web page is suspicious, legitimate or phishing.

The system structure comprises five components: (1) Website analysis and feature extraction; (2) an intelligent system; (3) a knowledge model that includes data gathered from PhishTank; (4) reported articles, projects and authentic sites; and (5) output process. The output process is a display of color-coded status with a text-based risk explanation made to inform the users. Red indicates suspicious sites and green indicates less severe websites.

Somesha et al. proposed a novel system using Deep Neural Network (DNN), Long Short-Term Memory (LSTM) and Convolution Neural Network (CNN) and using only ten features. The system architecture includes three modules: feature extraction, feature selection and classification. The Information Gain (IG) algorithm is used to select relevant features. This algorithm uses the ranking criterion to select the features by applying threshold. Finally, to assess the performance of the feature set, the feature set is trained and cross-validated against many different parameter combinations. The system uses DNN, LSTM and CNN-based models for phishing URL detection [14].

5.4.2 Solutions Using Deep Learning

Yang, Zhao and Zeng developed a novel approach to detect phishing web pages. This paper focuses more on features and they state that feature selection is the primary necessity for accurate detection [15]. They suggested a multidimensional attribute phishing detection technique based on a deep learning-based quick detection tool. In the CNN-LSTM step semantic and character dependency features of URL are captured and instant classification is performed. Preprocessing of data, feature extraction and classification are all included in this stage. Data preprocessing includes length normalization, uniform encoding and using an embedding layer to reduce the sparsity of the data. CNN is used to extract character sequence features from the supplied URL, which are then used by deep learning for fast classification. Context semantics and dependency features of URL character sequences are captured using the LSTM network. Soft max is used for quick classification. In the second step, multidimensional features-based URL statistical features, coding features and text features are considered. The result from the first step is then used by the XGBoost (eXtreme Gradient Boosting) ensemble learning algorithm, which has high classification accuracy. This approach has two ways of classification: quick classification using the softmax function of ratio of probabilities and the XGBoost classifier. This approach resulted in an accuracy of 98.99%.

5.5 Machine Learning as a Warhead

In this section we discuss how new technology like machine learning is used by the attackers as a tool to launch an attack. Researchers have made a study of the use of such techniques [16, 17]. The fake URLs that are created using artificial intelligence technology are called AI-based phishing URLs. In [18] the scientists proposed a system that generates effective AI-based phishing URLs after learning from the blacklisted URLs. This study first discusses various phishing detection strategies using Recurrent Neural Networks. They have proposed a system, DeepPhish, that can generate effective phishing URLs that cannot be detected by the various anti-phishing solutions that are currently used. In an overview of the system, the authors explored the data set of phishing URLs and identified a few threat actors. Threat actors are the domain names or few keywords that were efficient in bypassing the detection mechanisms. Using this data of threat actors, the LSTM model is trained to generate effective URLs. The dataset of 1.1M confirm phishing URLs was collected from PhishTank. Using this dataset on their own phishing detection system they identified the following threat actors:

Threat Actor 1: naylorantiques.com It was observed that this was a widely used domain name for launching an attack. Most common words were extracted from the URL path. Similar patterns were identified visually in the entire database. The authors recognized a total of 106 domains which Threat Actor 1 used for 1,007 attack URLs.

Threat Actor 2: vopus.org After exploring the data set, vopus.org was a commonly used domain and the pattern: tdcanadatrustindex.html was widely used for this domain. 102 URLs attacks were initiated using Threat Actor 2.

Threat Actor 3: creeksideshowstable.com The pattern recognized for this domain was the misspelling of the word "verification" i.e., "Paypal Virefication". 7,927 phishing URLs were created using this threat actor.

After identifying the threat actors, DeepPhish takes effective URLs as an input to the training AI algorithm. The LSTM Model is generated using these URL patterns as a one-hot encoding representation. Synthetic URLs are generated using the seed sentence that is used to predict the next character iteratively. The result of this experiment was that the success rate of Threat Actor 1 increased from 0.69% to 20.9% and Threat Actor 2 achieved an increase from 4.91% to 36.28%.

AlEroud and Karabatis implemented a system for sidestepping recognition of URL-based phishing attacks using GAN (generative adversarial deep neural networks) [19]. This approach has three major modules: (1) Generator (G) to produce adversarial URLs; (2) Discriminator (D) that takes these adversarial URLs and legitimate URLs as input and suggests weight updates to the Generator; and (3) the Phishing Detection Module (PD). Generator and Discriminator are neural networks in GAN where Generator aims to generate URLS that look real to Discriminator which distinguishes between the generated URLs and real ones. Discriminator acts as a tool to guide Generator to mislead the phishing detection system. The steps of the proposed system are:

1. Generating URL features from dataset of phishing and Genuine URLs.
2. Next step is Generator which is a feed forward neural network. The input to this module is feature vector, weights and the noise vector to produce adversarial URLs. Generator has three hidden layers with 120 neurons in each layer.
3. Discriminator: This module provides gradient information to Generator so that it is possible to update its weight values.

According to the results of the experiments, GAN is an effective technique for deceiving classifiers designed to defeat sophisticated attacks using URL-based features.

Researchers have also proposed techniques for generating phishing apps for android mobile devices using technologies like DL and image processing. We now discuss a technique proposed by Sen Chen et al. for generating phishing android apps called GUI-Squatting Attack. In this study the authors observed that successful phishing attack requires two conditions: page confusion and logic deception during attacks synthesis. This methodology has three phases (1) Component Extraction and Classification; (2) GUI Code Generation; and (3) Deceptive Code Generation. The overview of the system is such that a screenshot of a genuine app is captured followed by GUI component extraction from the image. Later, the researchers classify these components and generate the GUI code and then insert the deceptive code for stealing the personal information. This approach is as simple as creating only a single login page that looks exactly like original one and then extracting the confidential information of the user. In the deception code the authors have also implemented SSL/TLS authentication and user-identity verification via HTTPS connection for each phishing app to prevent detection by traffic analysis tools. Thus, a simple way of creating a login page gathering banking information and then redirecting the user to the genuine app is the methodology of this GUI-Squatting attack.

5.6 Comparative Analysis of the Techniques

In this section we compare the different systems discussed so far. We study their advantages and the future scope of the proposed systems (Table 5.1).

TABLE 5.1

Comparative Analysis of Detection Methodologies

Paper	Key Functionalities	Advantages	Future Scope
Machine Learning Techniques: Sameen, Han and Hwang. "PhishHaven—An Efficient Real-Time AI Phishing URLs Detection System." *IEEE Access* 8 (2020): 83425–83443.	HTML Encoding as a feature for phishing detections, Lexical analysis of URL-based heuristics, URL Hit Component and Modelics component where multiple ML algorithms are executed.	The system could detect tiny URLs as well as future AI-generated, phishing URLs with F1-measure of 98%,	The system can be improvised by using DL models. The Feature Selection Module also can be enhanced by using ML techniques.
Sahingoz et al. "Machine learning based phishing detection from URLs." *Expert Systems with Applications* 117 (2019): 345–357.	7 different classification algorithms are used, and NLP is used for feature selection.	Accuracy of system depends on selection of features and NLP-based features have better performance.	Feature extraction from tiny URLs can be added to improve the detection ration. The system must be tested for capability to detect AI-generated URLs.
Neural Network/Deep Learning: Verma et al. "Phishing Website detection using neural network and deep belief network." *Recent Findings in Intelligent Computing Techniques*. Springer, Singapore, 2019. 293–300.	ANN and DBN-based algorithms iare used for classification.	DBN approach gives better performance over other techniques.	Focus on AI-generated phishing URLs and a methodology for extracting the features of these AI-generated URLs can be designed.
Adebowale et al. "Intelligent web-phishing detection and protection scheme using integrated features of Images, frames and text." *Expert Systems with Applications* 115 (2019): 300–313.	An Adaptive Neuro-Fuzzy Inference System (ANFIS) and hybrid features like frames, images and text.	Users are warned with color-coded warning messages. Sugeno fuzzy model is used instead of Mamdani.	Intelligent system can be implemented at feature selection stage and evaluation must be done considering AI-generated URLs.

(Continued)

TABLE 5.1 (Continued)

Paper	Key Functionalities	Advantages	Future Scope
Yang, Zhao and Zeng. "Phishing website detection based on multidimensional features driven by deep learning." *IEEE Access* 7 (2019)	A multidimensional feature like URL statistical features, Web page code features, Web page text features used for phishing detection.	Focuses on Feature Selection, LSTM is used for feature extraction, Ensemble Machine learning algorithm gives higher accuracy in the results.	Deep Learning or Reinforcement Learning can be used for Feature Selection.
Somesha et al. "Efficient deep learning techniques for the detection of phishing websites." *Sādhanā* 45.1 (2020): 1–18.	Deep Neural Network (DNN), Long Short-Term Memory (LSTM) and Convolution Neural Network (CNN) using only ten features	Information Gain algorithm is used to select relevant features. Models resulted in accuracy of 99.52%, 99.57% and 99.43% for DNN, LSTM and CNN, respectively.	Real-time features like number of visits number of hits on the page can be evaluated. Features for the detection of embedded objects in the phishing website can be added.

The various techniques studied in this chapter are focused on detection algorithms and features used for the same. Datasets collected are mainly from PhishTank or other repositories. More research can be done on dataset collection and analysis of the dataset. The performance of the classifiers will vary if they are trained using an unbalanced dataset. Therefore, the researchers must also focus on achieving a balanced dataset for training the classification algorithms. Resampling methods like random under-sampling, random over-sampling, cluster-based over-sampling and informed over-sampling can be used to handle an imbalanced dataset [20, 21]. Imbalanced data can also be managed by altering the classification algorithms to detect the class accurately.

Any identification application requires an effective feature set for accurate recognition. Feature selection needs to be given equal importance with selection of classification of technique since it's the features that are essentially responsible for exact detection of phishing sites. Information Gain (IG), Gini Index (Gini), Chi-Square Metric (Chi-2), and Recursive Feature Elimination (RFE) can be used for feature ranking. In [14] the author has used IG for feature selection. More novel techniques can be designed for feature selection. This can result in identifying few but relevant features and eventually improve the response time of these applications for phishing detection.

5.7 Conclusion

In this chapter we studied various systems that are implemented for identifying phishing websites by applying latest technologies like ML, DL and NN. We have examined how for detection of the phishing URLs we need to focus on the feature selection stage of the system. We have encountered various novel features used by researchers like HTML encoding, NLP-based features and the number of visits to a web page. We can also conclude that researchers have not focused on this new area of cyber threat where ML is used as a tool to launch an attack. In Section 5.5. ML as a Warhead, we note that techniques like ML and GAN are used to generate these AI-based phishing URLs.

We would conclude with the need to draw our attention to these intelligent cyber-attacks and design a system that can detect these AI-generated phishing URLs.

References

[1] Lord, Nate. "Phishing Attack Prevention: How To Identify and Avoid Phishing Scams in 2019". October 6, 2020 [Online]. Available: https://digitalguardian.com/blog/phishing-attack-prevention-how-identify-avoid-phishing-scams

[2] Keelery, Sandhya. "Share of Consumers Using Smartphones for Digital Banking Activities across India as of January 2018, by Age Group". October 16, 2020 [Online]. Available: https://www.statista.com/statistics/870570/india-consumer-usage-of-smartphones-for-digital-banking-by-age-group/

[3] Singh, Charu. "Phishing Website Detection Based on Machine Learning: A Survey." *2020 6th International Conference on Advanced Computing and Communication Systems (ICACCS)*. IEEE, 2020.

[4] Bahnsen, Alejandro Correa, et al. "Deepphish: Simulating Malicious ai." *2018 APWG Symposium on Electronic Crime Research (eCrime)*. 2018.

[5] Sameen, Maria, Kyunghyun Han, and Seong Oun Hwang. "PhishHaven—An Efficient Real-Time AI Phishing URLs Detection System." *IEEE Access* 8 (2020): 83425–83443.

[6] Goswami, D. N., Manali Shukla, and Anshu Chaturvedi. "Phishing Detection Using Significant Feature Selection." *2020 IEEE 9th International Conference on Communication Systems and Network Technologies (CSNT)*. IEEE, 2020.

[7] Shekokar, N. M., C. Shah, M. Mahajan, and S. Rachh. "An Ideal Approach for Detection and Prevention of Phishing Attacks." *Procedia Computer Science* 49 (2015): 82–91.

[8] "What Is Machine Learning?" A Definition. Expert.ai [Online]. Available: https://www.expert.ai/blog/machine-learning-definition/

[9] Masurkar, Siddhesh, and Vipul Dalal. "Enhanced Lightweight Model for Detection of Phishing URL Using Machine Learning." *ICT Analysis and Applications*. Springer, Singapore, 2020. 45–56.

[10] Sahingoz, Ozgur Koray, et al. "Machine Learning Based Phishing Detection from URLs." *Expert Systems with Applications* 117 (2019): 345–357.

[11] A Beginner's Guide to Neural Networks and Deep Learning [Online]. Available: https://wiki.pathmind.com/neural-network

[12] Verma, Maneesh Kumar, et al. "Phishing Website Detection Using Neural Network and Deep Belief Network." *Recent Findings in Intelligent Computing Techniques*. Springer, Singapore, 2019. 293–300.

[13] Adebowale, Moruf A., et al. "Intelligent Web-Phishing Detection and Protection Scheme Using Integrated Features of Images, Frames and Text." *Expert Systems with Applications* 115 (2019): 300–313.

[14] Somesha, M., et al. "Efficient Deep Learning Techniques for the Detection of Phishing Websites." *Sādhanā* 45.1 (2020): 1–18.

[15] Yang, Peng, Guangzhen Zhao, and Peng Zeng. "Phishing Website Detection Based on Multidimensional Features Driven by Deep Learning." *IEEE Access* 7 (2019): 15196–15209.

[16] "AI Creates Phishing URLs That Can Beat Auto-Detection." GlobalNet [Online]. Available: https://www.gblnet.co.uk/news/ai-creates-phishing-urls-that-can-beat-auto-detection/

[17] Chen, Sen, et al. "Gui-Squatting Attack: Automated Generation of Android Phishing Apps." *IEEE Transactions on Dependable and Secure Computing* 18.6 (2019): 2551–2568. https://ieeexplore.ieee.org/stamp/stamp.jsp?arnumber=8913495

[18] Bahnsen, Alejandro Correa, et al. "Deepphish: Simulating Malicious AI." *2018 APWG Symposium on Electronic Crime Research (eCrime)*. 2018.

[19] AlEroud, Ahmed, and George Karabatis. "Bypassing Detection of URL-based Phishing Attacks Using Generative Adversarial Deep Neural Networks." *Proceedings of 2020 ACM International Workshop on Security and Privacy Analytics(IWSPA'20)*. March 18, 2020, New Orleans, LA, USA, ACM, New York, NY, USA. https://doi.org/10.1145/3375708.3380315

[20] Nagarhalli, Tatwadarshi P., Ashwini Save, and Narendra Shekokar. "Fundamental Models in Machine Learning and Deep Learning". *Design of Intelligent Applications using Machine Learning and Deep Learning Techniques*, edited By Ramchandra Sharad Mangrulkar, Antonis Michalas, Narendra Shekokar, Meera Narvekar, and Pallavi Vijay Chavan. Chapman and Hall/CRC, (2021): pp. 13–36.

[21] "Imbalanced Data: How to Handle Imbalanced Classification Problems" [Online]. Available at: https://www.analyticsvidhya.com/blog/2017/03/imbalanced-data-classification/

6

Analysis of Intelligent Techniques for Financial Fraud Detection

Narendra M. Shekokar
University of Mumbai, Mumbai, India

Aditi Vora
Dwarkadas J. Sanghvi College of Engineering, Mumbai, India

CONTENTS

6.1 Introduction

The emergence of intelligent techniques such as machine learning (ML) and deep learning (DL) has revolutionized prediction methods in sectors such as healthcare, crime and entertainment, among others. Machine learning comes from a branch of Artificial Intelligence (AI) that focuses on building applications that learn from the inputs given to the model, then the inputs are processed of inputs and finally given to the output function. There are many hidden layers in the model that does the processing. Machine learning learns from the data given. The amount of financial fraud is increasing day-by-day. A huge amount of data is stored [1–4]. The banks detect financial fraud to some extent but there are still some delays in detection. Recently there have been fraud cases detected in the State Bank of India, the Royal Bank of India, and others. Although, they have detected fraud to some extent, using intelligent techniques for detection is still in progress. Financial fraud detection causes stealing of important information and leads to damage of public property. Financial fraud occurs due to deliberate decisions and actions made by the people who handle money and other assets on behalf of employers or clients. Examples of financial fraud include the Ponzi scheme which is a fraudulent investing scam which generates return for earlier investors with the money taken from later investors. Using intelligent algorithms such as machine learning, we can detect various types of financial fraud and try to provide some solutions to prevent them. Machine learning is efficient because the data is in text format and it is comparatively easy to preprocess the data and perform training and testing on that dataset to try to find the output with maximum accuracy. Different algorithms are used by the authors such as SVM, and Decision Tree, among others [5–8]. Credit card fraud occurs when hackers try to gain personal information through stealing credit card details and gaining access to the owner's personal information. Credit card numbers can be stolen from an unsecured website. There are some tips to prevent credit card fraud such as never saving our credit card details on any unsecured site. The padlock icon shows that the website is secured and symbolizes a higher level of security. We should always check the security or encryption algorithm the website uses, and only then use the website. We should only buy products from a legitimate website. Research should be done on the company with which we want to make a transaction to check whether it is legitimate or not, and we should be cautious when we respond to special offers via email regarding loans, investments, and so on.

Credit card fraud detection is a concept in which fraudulent transactions are detected in our credit cards using various intelligent techniques such as machine learning, deep learning, artificial intelligence, and so on. Nonetheless, the challenges faced during detection are imbalanced data, enormous amounts of data being processed, as well as wrongly classified data, and all these reasons make fraud detection increasingly difficult.

Credit card fraud prevention can be achieved by using strong encryption algorithms, using two factor authentication and implementing firewalls to block the attackers. Figure 6.1 presents a graph that depicts frauds in banks.

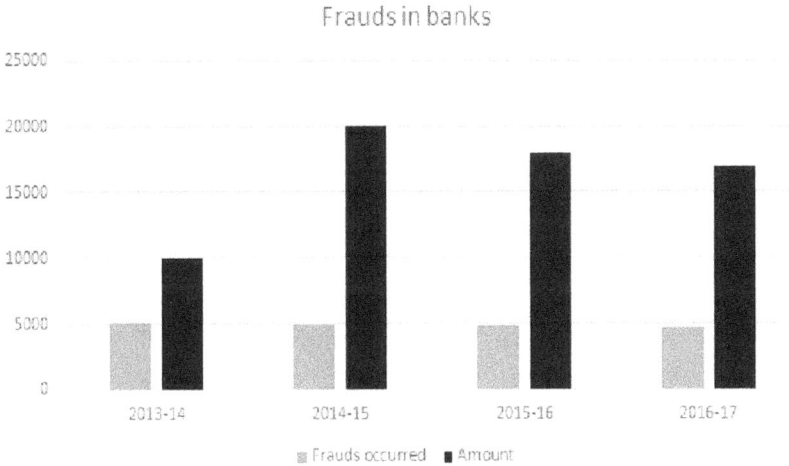

FIGURE 6.1
Frauds in banks.

As seen in Figure 6.1 depicting bank frauds and the amounts associated with the frauds, in 2013–2014 the most frauds occurred and then the number of frauds continued to slightly decrease until 2016–2017. The amount associated with the frauds, which is calculated in rupees, was at a maximum in 2014–2015.

6.2 Financial Fraud Detection Approaches

In this section, we discuss machine learning and deep learning approaches for financial fraud detection.

6.2.1 Machine Learning Approach for Financial Fraud Detection

Machine learning is an algorithm that comes under the category of artificial intelligence (AI). It has large datasets that are used in building models that can predict various behaviors. It takes input as the dataset from the input layer and then the processing and operations are completed in the hidden layer. The output layer produces the output after processing.

Machine learning algorithms learn from the data given to them and the accuracy can also be improved. In machine learning, algorithms are trained to find patterns and trends in the data. Machine learning can help in financial fraud detection to detect fraud patterns and behavior in the data and also to predict whether the transactions are fraudulent or not. Fraud detection mainly comes under classification problems.

Fraud detection using machine learning can be done using various algorithms such as the Support Vector Machine (SVM), Decision Trees, the Random Forest algorithm, Logistic Regression, the Naïve Bayes algorithm, among others. These algorithms provide a mechanism by which we can accurately predict fraudulent transactions. Machine learning algorithms are faster than other traditional ways of detecting fraud and can also deal with massive amounts of data.

Machine learning algorithms work by first collecting the data from various sources and analyzing the data, and then feature extraction is done by extracting the important features the system needs. In financial fraud detection, after collecting the data, preprocessing of the data is done to clean and filter the data so that the data becomes free from noise and is clean. The next step is feature extraction, in which features such as the source of the credit card or the location are extracted. The data obtained should always be more than required so that the model becomes more accurate to be used for prediction. After the feature extraction step, classifiers are applied such as the SVM, Naïve Bayes, logistic regression, and so on. The classifiers help to classify the transactions as fraudulent or non-fraudulent.

6.2.1.1 Challenges in the Machine Learning Approach

- *Overfitting and Underfitting*
  Overfitting and underfitting are the main problems in ML and they decrease the performance of the model. Overfitting occurs in supervised learning algorithms, and causes problems when our model tries to cover or fit all or more data points in the dataset. Because of this, the model gets inaccurate values and this reduces the performance of the model. Underfitting occurs when our model is not able to learn properly from the given training dataset, which reduces accuracy and also does not give reliable predictions.

- *Imbalanced Data*
  Imbalanced data means that the instance of one class is higher than the other, that is, the number of observations is not the same for all classes in the problem. The imbalanced dataset problem can be faced by binary as well as multiclass data. An imbalanced dataset can be controlled by using over-sampling, under-sampling and ensemble learning techniques.

- *Missing Data*
  Missing values in the dataset always affect the accuracy of the model. There are two approaches for removing the missing values from the dataset. The first approach is that we can remove all the missing values if the total percentage of missing values in the dataset is less than 5%. The second approach is that we can use a technique called imputation, if the total percentage of missing values in the dataset is more than 5%. Imputation is a technique in which the missing values in the dataset are replaced by mean, median or mode. However, we cannot always be dependent on the imputation technique, so the best way is to use Decision Trees for incomplete data.

6.2.1.2 Limitations of the Machine Learning Approach

- *Ethical Understanding*
  Machine learning does not have that much ethical understanding. In machine learning we trust the data given to us and the algorithms more than our own judgments. We cannot blame anyone if something goes wrong. If we take the example of driverless cars, if any driverless car kills someone on the road, then whose fault is it? This question arises in ethical understanding.

- *Diagnosis of the Error and Correction*
  When the model makes an error, correcting and rectifying it would be very time-consuming process and will require going through the complexities of the algorithms.

- *Time Constraints*
  We cannot make immediate and accurate predictions in machine learning since we have to travel through the historical data and the bigger the dataset, the longer it will take for processing.

6.2.2 Deep Learning Approach for Financial Fraud Detection

Deep learning is a category of machine learning algorithms which perform classification tasks directly from the images, audio, text, sound, and so on. There are large datasets in deep learning. Deep learning has multi-layered architecture in which the network learns from large amount of data. Deep learning algorithms perform calculations and predictions on the dataset and also improve accuracy by learning the behavior of the data. Deep learning does the analysis of data in real time.

Deep learning for fraud detection can be done using Convolutional Neural Network (CNN) algorithms, autoencoder-based clustering, Q learning, Recurrent Neural Network, Long Short Term Memory (LSTM) network, and so on. Deep learning algorithms have a hierarchy of non-linear transformation of inputs and give statistical output. They learn progressively more about the data as they go through different types of layers in the neural network.

6.2.2.1 Challenges in the Deep Learning Approach

- *Unstructured Data*
  Unstructured data does not follow a specified format. It is in the form of text, images, audio and video. It becomes difficult for prediction when using unstructured data.

- *Large Amounts of Data*
  Deep learning requires large datasets for processing. Large datasets are required to ensure that the machine produces the correct output. So, more parameters are also required for this, and it becomes a time-consuming process.

- *Optimization of Hyperparameters*
  Hyperparameters are those parameters whose value is predefined from the commencement of the learning process. If the values of the parameters are changed even by a small amount then this can invoke a huge change in the performance of the model.

- *High Performance Hardware*
  Deep learning requires large amounts of data. For solving real-world problems, the machine should have an adequate amount of processing power. For this reason, high performing GPUs are required.

6.2.2.2 Limitations of the Deep Learning Approach

- It lacks understanding about the exact output based on the input. On the basis of the training data given, we can only estimate the output but cannot assume that it would be exactly 100%. So, we must make different approximations.

- It lacks general intelligence. Human intelligence is due to connectivity and more accurate information. But, in the case of neural networks, if inaccurate information is fed in, then the outcomes could be wrong.

- Deep learning models cannot learn from a limited training set. Deep learning's intelligence is dependent on the training datasets used. If there are dynamically changing scenarios, then deep learning doesn't work.

- Deep learning is limited in its current form and cannot take complex decisions.

6.3 Literature Survey

6.3.1 Literature Survey for Machine Learning Approach

Rai et al. [1] proposed a method for credit card fraud detection using a neural network-based unsupervised learning algorithm along with autoencoder-based clustering, K-means, local outlier factor and isolation factor on a credit card system dataset, with the neural network algorithm having the highest accuracy at 99.87%. The advantage of this system was higher accuracy, and the disadvantage was that because an unsupervised algorithm was used, the data was not labeled so the processing time was longer.

Khatri et al. [2] proposed a method for credit card fraud detection using supervised machine learning algorithms and also provided comparison between different algorithms to distinguish between fraudulent and legitimate transactions. The algorithms used for prediction were Decision Tree, Random Forest, K-Nearest Neighbor (K-NN), Naïve Bayes and logistic regression. Sensitivity, precision and time were the various parameters for prediction. The best suited model was the Decision Tree for prediction but the sensitivity of K-NN was greater than the Decision Tree. Nonetheless, they concluded that the Decision Tree was the better option since the time taken for KNN was long. The disadvantage of this model was that since an imbalanced dataset was used, the dataset needed to be balanced.

Vidanelage et al. [3] proposed a system for fraud detection using multiple machine learning algorithms such as K-NN, Multilayer Perceptron (MLP), Gaussian Naïve Bayes (GNB), and Multinomial Naïve Bayes (MNB) on a synthesized dataset. The MLP algorithm gave the highest accuracy at 99.41%. Although K-NN and MLP gave almost the same accuracy, the MLP was preferred over the K-NN algorithm because the combined true positive and negative values of MLP were higher than K-NN. The disadvantage of this system was that although MLP performed well, its execution time was bit high.

Gyamfi et al. [4] proposed a system for fraud detection using the Support Vector Machine (SVM). The SVM performed better than the Back Propagation Network (BPN). A German and Australian credit card dataset was used here. The dataset was classified using SVM and trained using linear and logistic regression for anomaly detection. The SVM proved to be more reliable. The disadvantage was that the accuracy showed only the rate of true assignment and did not check whether it was true or false assignment.

Thennakoon et al. [5] proposed a system for real-time fraud detection using machine learning algorithms such as SVM, Naïve Bayes, K-NN, Logistic Regression, and time detection was performed. An imbalanced class was also handled in this system. The SVM gave the highest accuracy in this system at 91%. The disadvantage is that since it is a real-time monitoring system there could be some network issues.

6.3.2 Literature Survey for Deep Learning Approach

Zamini et al. [6] proposed a system for credit card fraud detection using autoencoder-based clustering which is a deep learning algorithm. An autoencoder with three hidden layers and K-means clustering was used here. A European dataset was used in this system. The accuracy of this model was 98.9%. The disadvantage is that the dataset was imbalanced because the proportion of fraud to nonfraud was 0.17%.

Bouchti et al. [7] proposed a system for credit card fraud detection using reinforcement learning. A Q-learning algorithm along with Convolutional Neural Network (CNN) was used here. Because of using CNN algorithm, the learning process was done directly from the raw input. The accuracy of the model was good. The disadvantage was that the agent had to deal with long range time dependencies and the future reward also had to be predicted.

Roy et al. [8] proposed a system for credit card fraud detection using deep learning algorithms such as Artificial Neural Network (ANN), Long Short-term Memory (LSTM), Gated Recurrent Unit (GRU) and Recurrent Neural Network (RNN). The LSTM and GRU algorithms outperformed the ANN and RNN algorithms. The disadvantage in this model was that since the dataset was imbalanced, sampling needed to be done.

Mubalaike et al. [9] proposed a system for intelligent financial fraud detection using an autoencoder algorithm and a Restricted Boltzmann Machine (RBM) algorithm. The RBM algorithm had the highest accuracy at 92.86%. The main disadvantage in this system was that although the total amount of transactions decreased, fraud remained at the same balance.

6.4 Proposed Model Architecture

The proposed model architecture is as shown in Figure 6.2.

The architecture description is given below:

1. Input Credit Card

   The first step in the fraud detection architecture is the credit card input is checked for whether the transactions are fraudulent or non-fraudulent.

2. Terminal Checking

   The second step in the fraud detection architecture is the terminal checking step. This means that all the terminal and important details regarding the credit card are checked. These details include: whether the entered credit card pin is correct or not; whether there is sufficient balance on the credit card or not; and whether the credit card in use is a blocked card or not. All these terminal check details are verified and if even one of the conditions is not satisfied from the terminal check, the card does not go for any further checks.

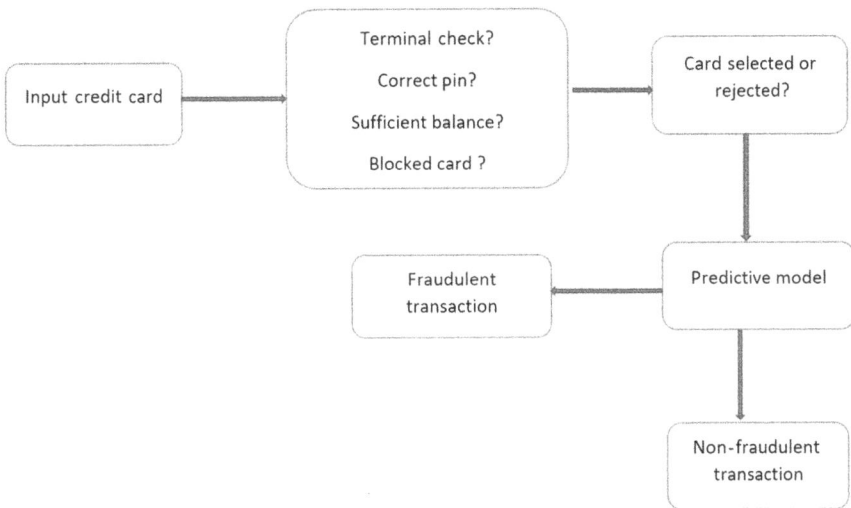

FIGURE 6.2
Fraud detection architecture.

3. Card Selected or Rejected
   The next step after the input of credit card and the terminal checks
   are done, is the card selected or rejected step. If all the three terminal
   check conditions are passed then the credit card is selected and if all the
   three conditions are not passed then the credit card is rejected for fraud
   detection.

4. Predictive Model
   The next step after the card selection or rejection step is the selection
   of a predictive model. The objective of a predictive model is to detect
   whether the transactions are fraudulent or not. The predictive model
   uses supervised machine learning algorithms to detect whether the
   transactions are fraudulent or not. There are different algorithms that
   can be chosen to select the correct predictive model. The algorithms are
   listed below:

   a. Decision Tree
      Decision Tree models are easy to interpret and simple to under-
      stand. A Decision Tree is a tree type of model structure which has
      different decisions and their consequences, and also the result out-
      comes. A Decision Tree has an internal node which represents a test
      on an attribute, and also has a branch which represents the outcome
      of the test attribute, while a leaf node represents the class label. The
      decision tree is represented as follows:

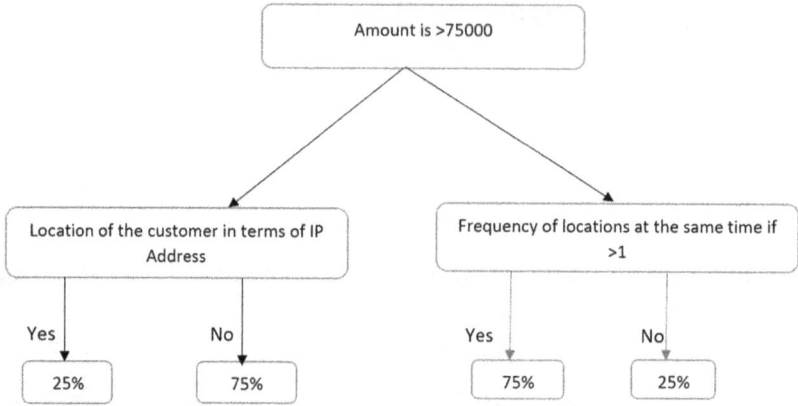

FIGURE 6.3
Fraud prediction using Decision Tree algorithm.

First, in the decision tree, we will check whether the transaction is greater than ₹75,000. If 'yes,' then we will check the location where the transaction is made. And if answer is 'no,' then we will check the frequency of the transaction. After that, as per the probabilities calculated for these conditions, we will predict the transaction as 'fraud' or 'non-fraud.' Here, if the amount is greater than ₹75,000 and the location is the same as that of the IP address of the customer, then there is only a 25% chance that the transaction is 'fraudulent', and a 75% chance is that the transaction is 'non-fraudulent.' Similarly, if the amount is greater than ₹75,000 and also the number of locations are greater than 1, then there is a 75% chance that the transaction is 'fraudulent' and a 25% chance that the transaction is 'non-fraudulent.' This is the way a Decision Tree in ML helps in creating fraud detection algorithms.

b. Random Forest Algorithm
Random Forest is a type of ML algorithm which uses a combination of different Decision Trees to improve the results. Each Decision Tree checks for different conditions and Decision Trees are trained on random datasets. The Random Forest algorithm can be used for both classification and regression problems. Random Forest is a classifier that contains a number of Decision Trees and finally takes an average to improve the accuracy of the dataset. The more trees we have, the better the accuracy we get in Random Forest. The Random Forest algorithm also takes less training time. A Random Forest example is shown below and in Figure 6.4:

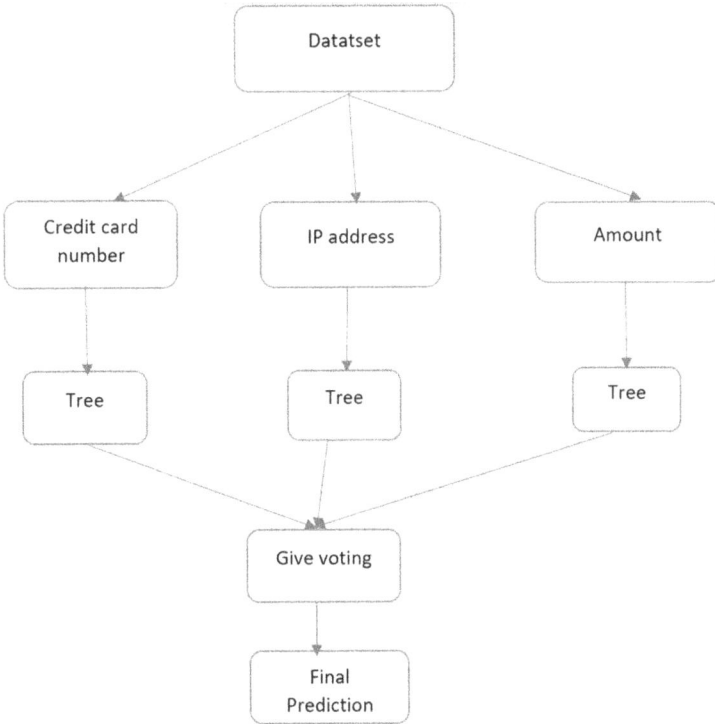

FIGURE 6.4
Fraud prediction using Random Forest algorithm.

When request for a transaction is given to the model, information like credit card number, IP Address and amount is checked. The data is then fed to the fraud detection algorithm. The fraud detection algorithm then selects different variables from the dataset and helps in splitting of the dataset into multiple Decision Trees. The subtrees contain conditions to check for an authorized transaction. After checking all the conditions, the trees give votes or probabilities for a transaction to be fraudulent or not fraudulent. Based on the combined result, the model will mark the transaction as fraudulent or not fraudulent.

c. Logistic Regression
Logistic Regression is a type of regression analysis. It is mainly beneficial for categorical data that gives the output as either 'yes' or 'no'. Logistic Regression helps data analysts make well-informed decisions. It also helps to minimize the loss. In Logistic Regression, instead of fitting a line we fit an "S" shaped curve that predicts the output as 'yes' or 'no'. Multinomial outcomes can also be predicted. A logistic regression curve is shown in Figure 6.5.

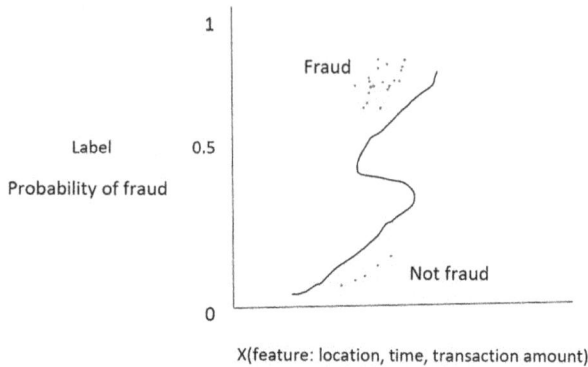

FIGURE 6.5
Fraud prediction using Logistic Regression.

6.5 Confusion Matrix

In machine learning and especially in statistical classification, a confusion matrix is also called an error matrix and has a table-like structure which tells us the performance of the algorithm in a tabular form and is specifically for supervised learning algorithms. Each row in the matrix represents the predicted class instances, and each column represents the actual class instances (or vice versa). All correct predictions are placed diagonally so that it becomes easy to check the table for prediction errors. A confusion matrix is not only used in binary classification problems but also in multi-classification problems. The confusion matrix is illustrated below:

N = Total predictions	Actual: No	Actual: Yes
Predicted: No	True Negative	False Positive
Predicted: Yes	False Negative	True Positive

6.6 Comparative Analysis of Decision Tree, SVM and Random Forest Algorithms

Table 6.1 presents a comparative analysis of the features of each of the three ML algorithms: Decision Tree, Logistic Regression and Random Forest.

TABLE 6.1

Comparative Analysis of Machine Learning Algorithms

Decision Tree	Logistic Regression	Random Forest
For fraud detection, Decision Tree considers only one tree at a time for building the output.	Logistic regression takes categorical values for fraud detection to classify the dataset and gives the possible probabilistic values, i.e., 0 or 1.	Random Forest is better than Decision Tree for fraud detection because it builds multiple decision trees to produce the output.
It is useful for classification and regression problems. Since fraud detection is a classification type problem, Decision Tree is the best option.	Since Logistic Regression is more beneficial for categorical problems, fraud cannot be detected with that much efficiency.	Since Random Forest algorithm is used for classification problems, fraud detection will be done efficiently.
Decision Tree is effective, easy for interpretation of results and suitable for structured data.	Logistic Regression is suited for structured data.	Random Forest does not depend highly on specific features and is also suitable for structured data.

6.7 Conclusion

With the increase in fraud in the financial sector, it is very important to maintain the security of organizations, to find effective methods to detect fraud and also find necessary solutions to prevent fraud. The fraud detection system architecture will help to detect fraud at each step and give feedback. Fraud detection systems using machine learning have proven to be effective in detecting fraud in the financial sector. This system can be proven more effective than other techniques used. In the near future, many other techniques could be used for fraud detection.

6.8 Future Scope

Machine learning and deep learning algorithms are well suited to predicting fraudulent transactions. Nonetheless, the major problems with machine and deep learning algorithms are that machine learning is suited for structured datasets that are not very large and deep learning works on image datasets. So, in our future scope we suggest more research on the availability of training sets that are large, so that we gain more accuracy in prediction and can train larger models. We should also look into other ways to implement deep learning in our model fuzzy systems, for outlier detection can also be combined with our network for better prediction. Fraudsters are becoming more intelligent day-by-day so to implement dynamic systems should be our utmost goal in the future.

References

[1] A. K. Rai and R. K. Dwivedi, "Fraud Detection in Credit Card Data Using Unsupervised Machine Learning Based Scheme," *2020 International Conference on Electronics and Sustainable Communication Systems (ICESC)*, Coimbatore, India, 2020, pp. 421–426, doi:10.1109/ICESC48915.2020.9155615.

[2] S. Khatri, A. Arora and A. P. Agrawal, "Supervised Machine Learning Algorithms for Credit Card Fraud Detection: A Comparison," *2020 10th International Conference on Cloud Computing, Data Science & Engineering (Confluence)*, Noida, India, 2020, pp. 680–683, doi: 10.1109/Confluence47617.2020.9057851.

[3] H. M. M. H. Vidanelage, T. Tasnavijitvong, P. Suwimonsatein and P. Meesad, "Study on Machine Learning Techniques with Conventional Tools for Payment Fraud Detection," *2019 11th International Conference on Information Technology and Electrical Engineering (ICITEE)*, Pattaya, Thailand, 2019, pp. 1–5, doi: 10.1109/ICITEED.2019.8929952.

[4] N. K. Gyamfi and J. Abdulai, "Bank Fraud Detection Using Support Vector Machine," *2018 IEEE 9th Annual Information Technology, Electronics and Mobile Communication Conference (IEMCON)*, Vancouver, BC, 2018, pp. 37–41, doi: 10.1109/IEMCON.2018.8614994.

[5] A. Thennakoon, C. Bhagyani, S. Premadasa, S. Mihiranga and N. Kuruwitaarachch, "Real-Time Credit Card Fraud Detection Using Machine Learning," *2019 9th International Conference on Cloud Computing, Data Science & Engineering (Confluence)*, Noida, India, 2019, pp. 488–493, doi: 10.1109/CONFLUENCE.2019.8776942.

[6] M. Zamini and G. Montazer, "Credit Card Fraud Detection Using Autoencoder Based Clustering," *2018 9th International Symposium on Telecommunications (IST)*, Tehran, Iran, 2018, pp. 486–491, doi: 10.1109/ISTEL.2018.8661129.

[7] A. E. Bouchti, A. Chakroun, H. Abbar and C. Okar, "Fraud Detection in Banking Using Deep Reinforcement Learning," *2017 Seventh International Conference on Innovative Computing Technology (INTECH)*, Luton, 2017, pp. 58–63, doi: 10.1109/INTECH.2017.8102446.

[8] A. Roy, J. Sun, R. Mahoney, L. Alonzi, S. Adams and P. Beling, "Deep Learning Detecting Fraud in Credit Card Transactions," *2018 Systems and Information Engineering Design Symposium (SIEDS)*, Charlottesville, VA, 2018, pp. 129–134, doi: 10.1109/SIEDS.2018.8374722.

[9] A. M. Mubalaike and E. Adali, "Deep Learning Approach for Intelligent Financial Fraud Detection System," *2018 3rd International Conference on Computer Science and Engineering (UBMK)*, Sarajevo, 2018, pp. 598–603, doi: 10.1109/UBMK.2018.8566574.

Further Reading

D. Tanouz, R. R. Subramanian, D. Eswar, G. V. P. Reddy, A. R. Kumar and C. V. N. M. Praneeth, "Credit Card Fraud Detection Using Machine Learning," *2021 5th International Conference on Intelligent Computing and Control Systems (ICICCS)*, 2021, pp. 967–972, doi: 10.1109/ICICCS51141.2021.9432308.

O. Adepoju, J. Wosowei, S. Lawte and H. Jaiman, "Comparative Evaluation of Credit Card Fraud Detection Using Machine Learning Techniques," *2019 Global Conference for Advancement in Technology (GCAT)*, 2019, pp. 1–6, doi: 10.1109/GCAT47503.2019.8978372.

Narendra Shekokar and Kriti Srivastiva, "Design of Machine Learning and Rule Based Access Control System with Respect to Adaptability and Genuineness of the Requester," *EAI Endorsed Transactions on Pervasive Health and Technology*, Issue 23, Sep 2020, doi: 10.4108/eai.24-9-2020.166359.

Narendra Shekokar and Vijay Shelke, "An Enhanced Approach for Privacy Preserving Record Linkage during Data Integration,"*6th IEEE International Conference on Information Management (ICIM)*, London, March 2020, pp. 27–29, doi: 10.1109/ICIM49319.2020.244689.

7

Evaluation of Learning Techniques for Intrusion Detection Systems

Tatwadarshi P. Nagarhalli, Ashwini M. Save and Narendra M. Shekokar
University of Mumbai, Mumbai, India

CONTENTS

7.1 Introduction

According to one estimate, as of July 2020 there are more than 4.8 billion internet users worldwide. Considering the fact that the world population is estimated at around 7.8 billion it can be understood that more than 60% of the citizens of this planet have access to the World Wide Web [1]. The year-on-year growth in the number internet users is more than 1000% [1].

This incredible worldwide penetration of the internet has come as a blessing for many people. There are many reasons for this exponential growth in the use of internet; one important reason, which has led people to readily embraced the internet, is the ease through which information can be assimilated and disseminated. Another reason is the simplicity of methods provided by financial institutions for undertaking financial transactions over the internet. In India alone, in the year 2020 more than $45 billion worth of transactions were completed digitally [2].

According to the US Federal Bureau of Investigation, individuals or the general public lost more than \$3.5 billion in internet fraud in the year 2019, an increase in excess of 29%. Also, more than 450,000 cases of internet fraud were reported with more than 32% increase in year-on-year figures for reported crimes [3]. Figure 7.1 shows the exponential rise in cybercrimes reported, while, Figure 7.2 shows the details of the losses incurred by individuals over a period of five years.

According to another report more than one third of Indian internet users are susceptible to some kind of malware attacks [4]. In fact, according to the National Cyber Security Coordinator, India is the most cyber-attacked country in the world [5]. And, considering the facts that more than 70% of these cyber-attacks are financially motivated [6] and a hacker attack is being reported every 39 seconds [7], it becomes imperative that measures are taken to safeguard personal information and data.

One important way through which cyber-attacks can be detected is the use of intrusion detection systems. Intrusion detection systems try to identify attacks on computer systems and alert the concerned authority responsible for dealing with cyber-attacks. These intrusion detection systems can be hardware-based systems or software applications designed to identify malicious activities. Generally, these detection systems monitor the computer systems or the network traffic for any malicious activities. If and when any malicious activities are detected, an alert is raised so that the attack can be tackled in an appropriate manner. This is the main aim of an intrusion detection system: to detect the attacks as soon as these attacks have been initiated. But, this also remains the biggest challenge.

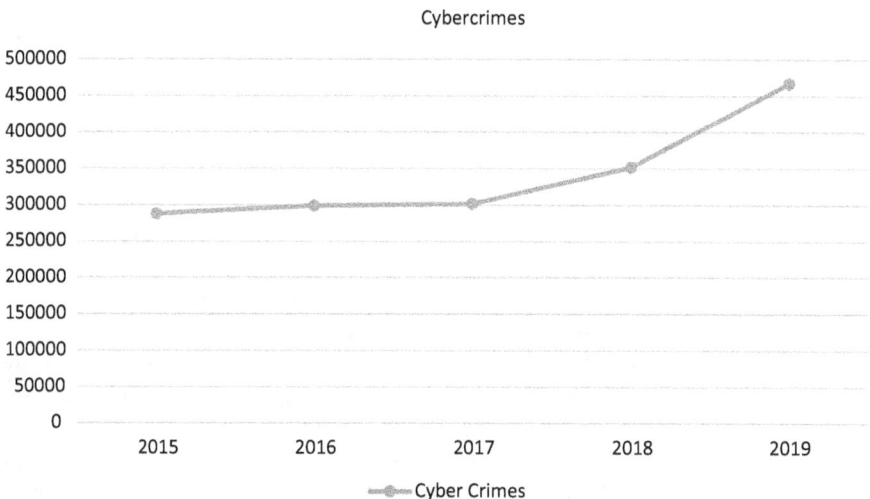

FIGURE 7.1
Cybercrimes reported.

Losses due to cybercrimes in $billion

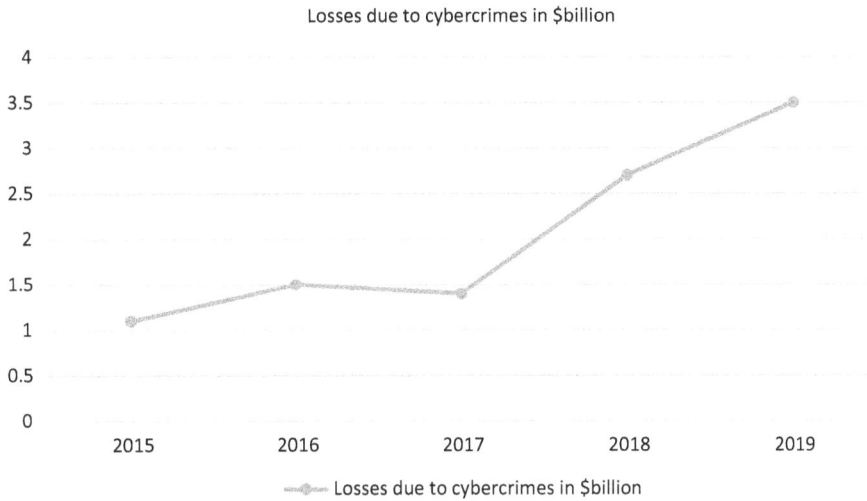

FIGURE 7.2
Losses incurred by individuals.

The detection and identification of cyber-attacks remain one of the most challenging aspects in cyber security because of the nature of the attacks. There are no fixed procedures which are followed by attackers; there are millions of ways in which a successful attack can be mounted. Over a period of time the intrusion detection systems have been upgraded and have tried to keep up with the challenges produced by the attackers. But, the attackers have been evolving at a much faster rate and in many cases they have been able to obfuscate the intrusion detection systems, and have been successful in mounting attacks on users. So, it becomes imperative that the intrusion detection systems also keep on evolving and tackling the challenges of cyber-attacks.

In order to keep pace with the advancing sophistication of cyber-attacks many new technologies have been incorporated into intrusion detection systems for better and more accurate detection of attacks. All the technologies used for intrusion detection can be grouped into two types: signature-based intrusion detection systems and anomaly-based intrusion detection systems [8].

In signature-based intrusion detection systems, a large database containing intrusion signatures is stored in the computing system [9]. The intrusion detection system tries to identify the patterns of the attacks and tries to map them with this signature database. If any of the patterns identified by the detection system matches with a pattern present in the attack signature database, then an alert is raised.

Signature-based intrusion detection systems are very easy to develop and have proved to be highly accurate when the attack pattern has been mentioned in the stored database [10]. But, on the flip side, signature-based intrusion detection systems can detect only previously known attacks. So, it is almost impossible for a signature-based intrusion detection system to identify new attacks whose signatures are not present in the attack signature database. As the attacks evolve and new attack signatures are identified, these new attack signatures must be loaded into the attack signature database. Because of this, the size of the signature database keeps on increasing as well. Moreover, owing to the increased database size, the intrusion detection system might take more time to detect attacks, thus defeating the very purpose of detecting intrusions.

In anomaly-based intrusion detection systems, the intrusion attacks are identified by checking for any abnormalities [11]. If any abnormalities are detected, then alerts are raised, and competent authorities are notified. The anomaly-based intrusion detection system themselves can be categorized into three types: (1) statistic-based systems, (2) knowledge-based systems and (3) machine learning-based systems [8]. Figure 7.3 shows the classification of intrusion detection systems.

Statistics-based intrusion detection systems make use of statistical methods for finding the anomalies in the system or in the network [12]. Different statistical methods, like time series models, have been used for identifying anomalies. Alternatively, knowledge-based intrusion detection systems make use of knowledge bases. Generally, it has been observed that expert systems are designed for the detection of intrusions [13]. Expert systems are systems designed for a specific purpose which can emulate human decision making. The effectiveness of knowledge-based intrusion detection depends on the knowledge base created for detection the intrusions [14].

Machine learning is a subfield of artificial intelligence which endeavors to learn from past experiences [15]. In recent times, machine learning has played a very important role in many major fields, including security, and has provided a very positive impact. There are many different techniques in machine learning which have been investigated for detection of intrusions based on anomalies. Comparatively, machine learning techniques have shown promising results. Deep learning techniques, which are specialization of machine learning [16], have also been tested for better anomaly-based detection of intrusion. Machine learning techniques have shown great promise in the detection of anomalies and intrusion, especially considering the dynamically evolving field of cyber security.

So, the chapter proposes to perform a detailed study of five learning techniques: Logistic Regression, Decision Tree, Artificial Neural Network, Random Forest and Gradient Boosting, and evaluate their effectiveness in the detection of intrusions. The performance of a machine learning technique depends on various factors like the way the data is provided, namely categorical or continuous, and this aspect has been considered as well. So, these machine learning techniques have been analyzed and evaluated on four parameters: (1) the type of input data given to the model; (2) the number of

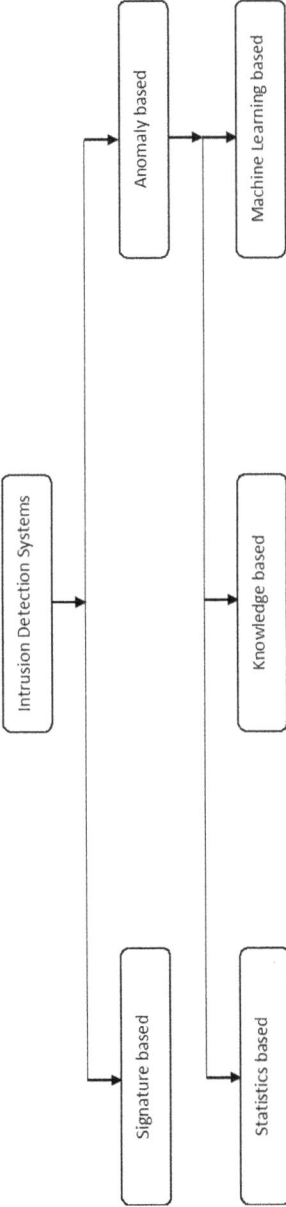

FIGURE 7.3
Classification of intrusion detection systems.

classes present; (3) the number of training instances present for each class; and (4) the number of features in the training data.

Apart from evaluating the effectiveness of the machine learning techniques, the chapter proposes to give reasons for the effectiveness and ineffectiveness of any technique while performing the detection of intrusions.

In the following sections, a brief review of existing literature is presented. Then, the intrusion detection system dataset used for the evaluation of machine learning techniques is elaborated. After that, a detailed analysis and evaluation of different machine learning techniques on the identified datasets is carried out. Finally, concluding remarks are made.

7.2 Review of Literature

Machine learning techniques have enjoyed many successes in various fields. This has prompted further research and employment of these learning techniques in very diverse fields. One such field of study where machine learning techniques have showed great promise is anomaly-based intrusion detection systems.

Many different machine learning techniques have been trialed in order to find the best technique which can identify the anomaly. For example, Halimaa and Sundarakantham [17] study and compare machine learning techniques with Naïve Bayes and Support Vector Machine for intrusion detection. In their opinion, comparatively, the Support Vector Machine performs better.

However, Gautam and Doegar [18] propose an ensemble approach. Their paper proposes to perform experiments on the KDDCup99 dataset by applying Naïve Bayes, Adaptive Boost and Partial Decision Tree algorithms in parallel in order to increase the accuracy of the system.

On the other hand, Park et al. [19] performed experimentations for anomaly detection on the Kyoto dataset. The Random Forest algorithm was used for the experimentations. Furthermore, Yihunie et al. [20] experimented with the Random Forest algorithm on the NSL-KDD dataset.

There have been evaluation papers as well which have evaluated the accuracy of different machine learning techniques. Ahmad et al. [21] performed comparisons of Support Vector Machine, Random Forest and a proposed extreme learning machine. The experimentations were performed on the NSL-KDD dataset.

In addition to supervised learning approaches, unsupervised learning approaches have also been employed in order to get good anomaly detection accuracy. But, generally, these unsupervised learning approaches have been used in combination with supervised learning approaches. Both Zhanga et al.

[22] and Liang et al. [23] propose a hybrid approach of using K-means clustering and Support Vector Machine for anomaly detection. The only difference is that paper [22] uses BIT19 dataset, while paper [23] proposes to work with the NSL-KDD dataset.

It can be seen that there have been quite a number of research works on the identification of attacks in anomaly-based intrusion detection systems using machine learning approaches. But, it can be observed that there are very few works which have carried out detailed evaluation and analysis of modern machine learning approaches. It is believed that a detailed evaluation and analysis of modern machine learning techniques would help in the deployment of an effective technique which would identify any anomalous behavior.

7.3 Dataset

For machine learning techniques to work, they have to be trained. It can also be said that the efficiency of a machine learning technique depends largely on the dataset provided to it for training. So, a dataset plays a very important role in determining the overall performance of a machine learning model.

For intrusion detection systems, there are two very important and popular datasets. These datasets are the KDDCup99 dataset [24] and the NSL-KDD dataset [25]. The NSL-KDD dataset is a derivative of the KDDCup99 dataset. Further, the KDDCup99 dataset itself has been derived from a survey conducted by the DARPA Intrusion Detection Evaluation Program, managed by the Massachusetts Institute of Technology's Lincoln Lab.

It was observed that the KDDCup99 dataset contained much redundant data [26]. That is, in this dataset many data entries were repeated. So, a new dataset was prepared where the redundancies were removed, while the features or attributes of the KDDCup99 dataset were retained. This new dataset was called the NSL-KDD dataset. Considering the popularity of both these datasets, the evaluation of important machine learning techniques has been carried out on both these datasets.

Both the KDDCup99 dataset and NSL-KDD dataset contain a total of 41 features and 23 classes. Of these 23 classes, 1 is normal and the remaining 22 are different kind of attacks. The KDDCup99 dataset contains a total of 494,020 data entries, while the NSL-KDD dataset contains a total of 125,973 data entries.

Details of the distribution of the 494,020 instances of the KDDCup99 dataset and the distribution of the 125,973 instances of the NSL-KDD dataset are illustrated in Table 7.1.

TABLE 7.1

Number of Instances per Class

Sr. No.	Name of the Class	Number of Instances in KDDCup99	Number of Instances in NSL-KDD
1.	Normal	97,277	67,343
2.	Buffer Overflow	30	30
3.	Load Module	9	9
4.	Perl	3	3
5.	Neptune	107,201	41,214
6.	Smurf	280,790	2,646
7.	Guess Password	53	53
8.	POD	264	201
9.	Teardrop	979	892
10.	Port Sweep	1,040	2,931
11.	IP Sweep	1,247	3,599
12.	Land	21	18
13.	FTP Write	8	8
14.	Back	2,203	956
15.	IMap	12	11
16.	Satan	1,589	3,633
17.	PHF	4	4
18.	NMap	231	1,493
19.	Multihop	7	7
20.	Warezmaster	20	20
21.	Warezclient	1,020	890
22.	Spy	2	2
23.	Rootkit	10	10
Total		494,020	125,973

It can be observed from Table 7.1, how imbalanced both these datasets are. For the KDDCup99 Dataset, the Smurf class has 280,790 data instances, yet the Spy class has only two data instances. Similarly, even for the NSL-KDD dataset, the Normal class has a total of 67,343 instances whereas, the Spy class again has only two instances.

Further, these 22 attacks or classes have been categorized into four attacks. These four categories are the Denial of Service attack (DOS), Probing attack (Probe), User to Root attack (U2R) and Remote to Local attack (R2L). Figure 7.4 shows the details of the categorization of the 22 attacks into the four categories.

Table 7.2 shows the number of data instances available for each of the 4 attacks after the categorization.

It can be observed that, even after consolidation of 22 classes into four attack types, the datasets remain highly imbalanced. For the KDDCup99

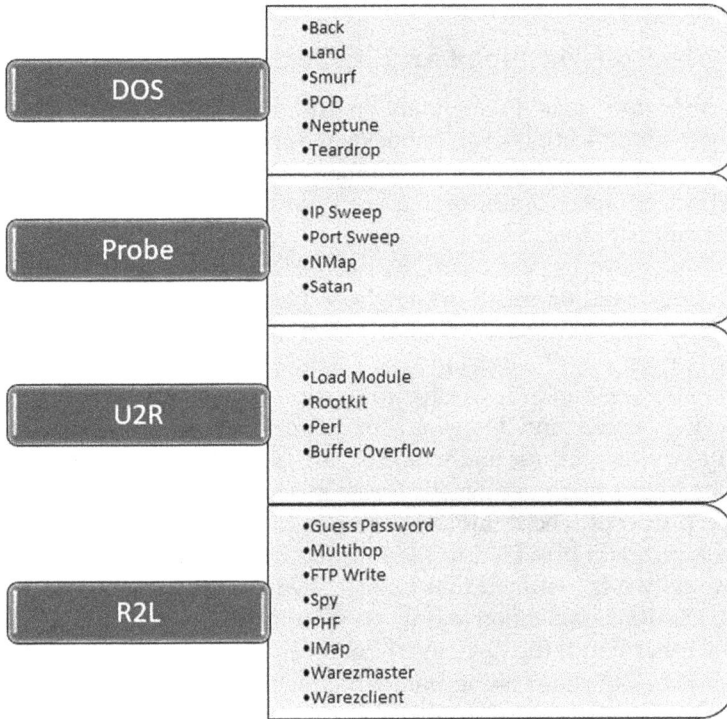

DOS
- •Back
- •Land
- •Smurf
- •POD
- •Neptune
- •Teardrop

Probe
- •IP Sweep
- •Port Sweep
- •NMap
- •Satan

U2R
- •Load Module
- •Rootkit
- •Perl
- •Buffer Overflow

R2L
- •Guess Password
- •Multihop
- •FTP Write
- •Spy
- •PHF
- •IMap
- •Warezmaster
- •Warezclient

FIGURE 7.4
Categorization of the attacks.

TABLE 7.2

Number of Data Instances in Different Attack Categories

Sr. No.	Name of the Attack	Number of Instances in KDDCup99	Number of Instances in NSL-KDD
1.	Normal	97,277	67,343
2.	DOS	391,458	45,927
3.	Probe	4,107	11,656
4.	U2R	52	52
5.	R2L	1,126	995
Total		494,020	125,973

dataset, the DOS attack has the highest number of instances with 391,458 and the the U2R attack has the lowest number of instances with 52 entries. For the NSL-KDD dataset, the highest number of instances have been recorded for the Normal class with 67,343 data entries and the U2R attack with 52 instances has the lowest data entry figure.

7.4 Analysis of Machine Learning Techniques

As an important task in anomaly-based intrusion detection systems is the proper identification of the attacks, a supervised learning approach is required. Also, as both the identified datasets have different classes of attacks and distinct groups, classification algorithms are required under the supervised learning approaches.

For the purpose of evaluation, five classification machine learning techniques have been identified which have been trained and tested on the KDDCup99 and NSL-KDD datasets. The identified techniques are: (1) Logistic Regression, (2) Decision Tree, (3) Simple Artificial Neural Network, (4) Random Forest and (5) Gradient Boosting [27]. Of these five techniques Logistic Regression and Decision Tree algorithms are legacy techniques, while the remaining three methods are state-of-the-art techniques.

Logistic Regression uses logistic function or the sigmoid function for classification purposes. The sigmoid function is an 'S' curve and using this function the input data provided to the Logistic Regression is classified. Due to this, generally it is assumed that Logistic Regression works best when performing binary classification and when working with numerical input data.

On the other hand, the Decision Tree develops a tree-like structure based on the features of the training data which is utilized for classification purposes. For the construction of the tree, the concept of information gain is used. In most general terms trained decision trees are simple 'if-then-else' statements used for classification. Decision Trees can work with both numerical and categorical data.

The Artificial Neural Network tries to mimic the functionality of the human neuron of collaborative learning. This collaboration artificial neural network is possible with the help of weights and biases associated with neurons. Initially these weights and biases are assigned at random. Using a back propagation algorithm, these weights and biases are recalculated in order to perform classification. Because of the presence of these weights and biases, artificial neural network work is numerical data. Deep learning techniques are based on the simple Artificial Neural Network.

Random Forest and Gradient Boosting are both ensemble approaches. That is, they develop more than one decision structures. Random Forest algorithms develop a number of decision tree structures using subsets of features given in the training data. All these different decision trees constructed by the Random Forest algorithm perform classification of the dataset independently, and the class chosen by the maximum number of decision trees is chosen as the answer of the Random Forest algorithm.

Similar to the Random Forest approach, the Gradient Boosting algorithm also makes use of multiple decision tree structures. While the Random Forest approach develops a number of decision trees at random, Gradient Boosting algorithms develops one decision tree at a time. The structure of a decision tree and the features selected for the construction of a particular decision tree depends on the classification results produced by the immediate previous decision tree. As both Random Forest and Gradient Boosting algorithms create multiple decision trees, both these techniques can handle numerical as well as categorical data.

A number of factors determine the performance and the efficiency of machine learning algorithms. Some of the important factors that affect the performance of these learning techniques are: (1) the type of input data given to the model, (2) the number of classes present, (3) the number of training instances present for each class and (4) the number of features in the training data. The evaluation and analysis of the effectiveness of the four identified learning techniques has been carried out while considering these parameters.

7.4.1 Input Data

There are two types of input data that can be given to a machine leaning algorithm: categorical data and continuous numerical. Categorical data means the feature values of the data are distinct groups, like male and female, whereas, continuous numerical data are finite numerical values, like 1, 2, and so on.

While techniques like Logistic Regression and Artificial Neural Network work best when the given data is continuous numerical, Decision Tree, Random Forest and Gradient Boosting can work with, and handle, both continuous and categorical input data.

The KDDCup99 and NSL-KDD datasets contain a mixture of both these types of input data. Of the 41 features in the dataset, seven features are categorical in nature and the remaining are continuous numerical values. It is very difficult for a machine learning algorithm to handle input data which contains a mixture of continuous numerical and categorical data.

So, the accuracy of the machine learning techniques were analyzed in two ways. First, by giving the dataset, as seen, to the five machine learning approaches. In this case the input data was considered to be continuous numerical values. In the second case all the 34 continuous features were converted into categorical features containing either three or four categories based on their range of values.

For example, the values for the feature named 'src_bytres' range from ten to hundreds and to thousands, so, these values were converted into four categories. On the other hand, the feature named 'duration' has much less variation, hence, these values were converted into three categories.

The different python codes developed for evaluation can be accessed through GitHub [28]. The python file names have been mentioned in the tables.

The accuracy produced under both these cases by the machine learning techniques when KDDCup99 and NSL-KDD datasets were provided to it separately, has been shown in Table 7.3.

TABLE 7.3

Accuracy for Numerical and Categorical Input Data

Sr. No.	Data Type (GitHub Python File Name)	Algorithm Name	KDD Cup99 Result in %	NSL-KDD Result in %
1.	Numerical (IDS)	Logistic Regression	99.18	83.57
		Decision Tree	99.97	99.74
		Artificial Neural Network	56.75	0.03
		Random Forest	99.97	99.82
		Gradient Boosting	99.85	98.77
2.	Categorical (IDS-Cat)	Logistic Regression	99.33	95.06
		Decision Tree	99.49	98.27
		Artificial Neural Network	98.40	88.17
		Random Forest	99.50	98.33
		Gradient Boosting	99.30	96.46

TABLE 7.4

Accuracy for Numerical and Categorical Input Data with Five Classes

Sr. No.	Data Type (GitHub Python File Name)	Algorithm Name	KDD Cup99 Result in %	NSL-KDD Result in %
1.	Numerical (IDS-Bal)	Logistic Regression	99.17	84.84
		Decision Tree	99.97	99.75
		Artificial Neural Network	94.32	22.93
		Random Forest	99.98	99.87
		Gradient Boosting	99.83	98.44
2.	Categorical (IDS-Bal2)	Logistic Regression	99.22	94.50
		Decision Tree	99.49	98.45
		Artificial Neural Network	79.13	91.13
		Random Forest	99.51	98.48
		Gradient Boosting	99.23	95.52

From Table 7.4 it can be seen that for numerical data, with the exception of the Artificial Neural Network, all the techniques produced very good results for both the datasets. Moreover, when categorical data was provided, all five techniques produced very good results for both datasets.

Even though the results seem impressive certain things need to be considered. For training and testing with numerical continuous data the original dataset was kept untouched so, it is unknown how the categorical data was treated by the different machine learning techniques. Also, when the dataset was converted into categorical, even then Logistic Regression, which is considered to work well with numerical data, gives an accuracy of more than 99%. This means that even the categorical data was considered as numerical. So, it would be better to perform more detailed analysis.

7.4.2 Number of Classes

It is generally observed that the performance of some machine learning techniques degrade with the increase in the number output classes. So, originally, with around 23 classes in the datasets, the performance of the machine learning techniques could be undesirable and also untraceable. Untraceable because, with the increase in the number of classes it becomes difficult to understand whether the model is performing as expected or not.

So, the number of classes was brought down to five. This was using details from Figure 7.2 and Table 7.3. Now, instead of having 23 classes, the dataset would now contain only five classes, including normal class. The accuracy results produced on both the datasets are detailed in Table 7.4.

From Table 7.5 it can be inferred that Decision Tree, Random Forest and Gradient Boosting techniques produced good results in all the above cases. The performance of Artificial Neural Network increases drastically from 22.93% for numerical data to 91.13% to categorical data when the NSL-KDD dataset was provided.

TABLE 7.5

Accuracy Produced after Random Oversampling

Sr. No.	Data Type (GitHub Python File Name)	Algorithm Name	KDD Cup99 Result in %	NSL-KDD Result in %
1.	Numerical (IDS-O-N)	Logistic Regression	38.53 (Iteration Reached Limit)	27.71 (Iteration Reached Limit)
		Decision Tree	99.99	99.93
		Artificial Neural Network	22.59 (Abnormal Termination)	21.32 (Abnormal termination)
		Random Forest	99.99	99.98
		Gradient Boosting	97.5	95.19
2.	Categorical (IDS-O-C)	Logistic Regression	91.55 (Iteration Reached Limit)	89.29 (Iteration Reached Limit)
		Decision Tree	96.79	96.08
		Artificial Neural Network	20 (Abnormal termination)	76.59 (Abnormal termination)
		Random Forest	96.79	96.10
		Gradient Boosting	91.50	87.89

This is again undesirable because, an input to a neural network has to be a numerical value. So, it has to be concluded that the neural network might be treating the categorical data which has been encoded as 0, 1, and so on, as a numerical value of 0, 1, and so on. Working with these three to four numbers would be simpler than working with diverse real values which range from tens to thousands. This might be one of the reasons why there is a spike in the accuracy of the neural network.

7.4.3 Number of Training Instances for Each Class

Another important factor that affects the overall accuracy and performance of a machine learning model is the number of data instances for each class in the training data. The reason for this is, a class with higher number of data instances is studied better by a learning model than a class with a lower number of data instances.

If there is a huge disparity of data instances between the classes, then the results cannot be trusted either. As already mentioned, both the KDDCup99 and NSL-KDD datasets are highly imbalanced. The imbalance in the dataset continues even after consolidation of the classes after categorization.

Generally, there are two ways of dealing with imbalance of the dataset in machine learning: oversampling and undersampling [29]. The analysis of machine learning techniques after oversampling and undersampling has been conducted on KDDCup99 and NSL-KDD datasets after the categorization of the datasets, that is, sampling has been carried out on the datasets with five classes.

7.4.3.1 Oversampling

Oversampling is a technique of creating or replicating multiple instances of a class with a lower number of data instances (also called a minority class) in an imbalanced dataset. In a dataset, the class with the highest number of data instances is called the majority class and all the other classes are minority classes. Replication of data is carried out for all the minority classes till the number of data instances reach the same number as that of the majority class.

From Table 7.3 it can be observed that, for the KDDCup99 dataset, the DOS attack has the highest number of instances hence, it is the majority class with 391,458 data instances and the remaining four classes are the minority classes. Replication of data is carried out for all the four minority classes till the number of data instances for each of these classes reaches 391,458. So, after oversampling, all the five classes in the dataset contain 391,458 data instances each.

There are two simple and effective ways of conducting oversampling: Random Oversampling and Synthetic Minority Over-sampling Technique (SMOTE) [30]. In Random Oversampling, replication of data instances of the minority class is carried out randomly. Whereas in SMOTE, the replication of data in the minority class is carried out while considering the similarity between the data instances within the minority class [31]. That is, instead of performing replication at random, the SMOTE considers the feature space of the minority class input data. Using this feature space, semantic similarity and characteristics of the input data points of these classes are identified. Using this information, the data points in the minority classes are generated. As the SMOTE has a better understanding of the data points within the minority class, the data points generated using this technique are comparatively considered to be more authentic and better for evaluation of a machine learning model.

Table 7.5 shows the accuracy results produced by the machine learning techniques after Random Oversampling has been conducted.

While Random Oversampling can be carried out for both numerical and categorical data, SMOTE oversampling can be carried out only on numerical valued data. Table 7.6 shows the accuracy results produced by the machine learning techniques after Oversampling was carried out using the SMOTE technique.

From Tables 7.6 and 7.7 there are a few important observations which can be made. First, it can be observed that there are hardly any differences or changes in accuracies for both types of oversampling methods, for both the datasets. So, it can be said that for the two datasets under consideration, the oversampled data produced by Random Oversampling and SMOTE oversampling should be quite similar.

Another observation which can be made is, the Logistic Regression and Artificial Neural Network methods were terminated abruptly. Here, it has to be noted that after oversampling the KDDCup99 dataset contained a total of 1,957,290 data instances and the NSL-KDD dataset contained a total of 336,715 data instances. So, it has to be said that both these machine learning methods could not handle the huge number of data instances created by oversampling of the data.

TABLE 7.6

Accuracy Produced after SMOTE Oversampling

Sr. No.	Data Type (GitHub Python File Name)	Algorithm Name	KDD Cup99 Result in %	NSL-KDD Result in %
1.	Numerical (IDS-OS-N)	Logistic Regression	37.61 (Iteration Reached Limit)	23.85 (Iteration Reached Limit)
		Decision Tree	99.87	99.85
		Artificial Neural Network	27.0 (Abnormal Termination)	21.24 (Abnormal Termination)
		Random Forest	99.99	99.97
		Gradient Boosting	97.19	95.09

TABLE 7.7

Accuracy Produced after Random Undersampling

Sr. No.	Data Type (GitHub Python File Name)	Algorithm Name	KDD Cup99 Result in %	NSL-KDD Result in %
1.	Numerical (IDS-U-N)	Logistic Regression	73.08	48.08
		Decision Tree	92.31	80.77
		Artificial Neural Network	17.31	21.15
		Random Forest	98.08	90.38
		Gradient Boosting	94.23	86.54
2.	Categorical (IDS-U-C)	Logistic Regression	82.69	82.69
		Decision Tree	84.61	86.54
		Artificial Neural Network	15.13	63.46
		Random Forest	88.46	88.46
		Gradient Boosting	90.38	80.77

7.4.3.2 Undersampling

In undersampling, first and foremost the minority class with the least number of data instances is identified, and is called the least minority class. Then, the data instances of all the other classes are deleted in order to match the number of data instances of the least minority class. At the end of the undersampling process all the classes have the same number of data instances as the least minority class.

From Table 7.3 it can be observed that, in both the KDDCup99 and NSL-KDD datasets, the User to Root attack class has the lowest number of data instances with 52 entries. So, after undersampling all the remaining classes also have only 52 data instances.

Table 7.7 shows the accuracy results of the machine learning techniques on KDDCup99 and NSL-KDD datasets after Random Undersampling has been carried. Random Undersampling is a technique where the data instances of all the non-least minority classes are deleted at random.

After observing Table 7.8 one may feel satisfied with the results but, making decisions based on these results may not be wise. This is especially true of the two datasets under consideration. The main reason for this is the level by which the classes are all represented.

In the NSL-KDD dataset the Normal class with 67,343 data instances is the majority class. But, after undersampling it has only 52 instances that is, only 0.077% of the original data has been represented. For the KDDCup99 this ratio is even worse. In the KDDCup99 dataset the DOS attack class with

TABLE 7.8

Accuracy Produced after PCA ($n = 2$) on Datasets with 23 Classes

Sr. No.	Data Type (GitHub Python File Name)	Algorithm Name	KDD Cup99 Result in %	NSL-KDD Result in %
1.	Numerical (IDS-N-PCA)	Logistic Regression	86.57 (Iteration Reached Limit)	44.29 (Iteration Reached Limit)
		Decision Tree	99.81	92.98
		Artificial Neural Network	90.48 (Abnormal Termination)	53.29 (Abnormal termination)
		Random Forest	99.83	93.01
		Gradient Boosting	99.31	92.37

391,458 data instances is the majority class, of which only 52 instances are considered for training and testing, which is only 0.01% of the original data.

Hence, it has to be said that the machine learning methods would have insufficient data for training and testing. So, the results produced by the machine learning techniques under such conditions are neither optimum nor complete.

7.4.4 Number of Features

The number of features also plays a very important role in determining the overall efficiency of a machine learning algorithm. With the increase in the number of features, there is an exponential increase in the complexity of machine learning techniques [32] which might adversely impact the outcome of a system.

For the datasets under consideration, there are 41 features. Working with this many features not only increases the complexity of the machine learning techniques, but also raises questions over the efficiency of the techniques. So, reducing the number of features is an important task.

One of the popular techniques used for reducing the number of features in a dataset is principal component analysis (PCA) [33]. In PCA, important components are computed and identified which would replace the original features and feature values.

Evaluation of machine learning techniques has been carried out by applying principal component analysis on both the datasets for all variants of the data. That is, principal component analysis has been applied on the original data with 23 classes, on the categorized data with five classes, and on both types of oversampled data. Principal component analysis has not been applied on the undersampled data because, as it has already been established, for the dataset under consideration undersampled data is not appropriately representing the original dataset.

After principal component analysis has been carried out, machine learning techniques have been implemented in order to find the accuracy scores. For testing purposes, the number of principal components has been kept at two. That is, the machine learning models will have only two features to work with. Also, an important thing to note here is that principal component analysis can only be conducted on continuous numerical data. So, the accuracy results have been obtained only on the numerical data and not on categorical data.

Table 7.8 shows the accuracy results produced by the machine learning techniques on the dataset with 23 classes and after converting 41 features into two principal components.

Accuracy results by machine learning techniques on five classes of attack categories after principal component analysis has been carried with two principal components has been detailed in Table 7.9.

Table 7.10 shows the accuracy results on Random Oversampled data with five classes and two principal components.

TABLE 7.9

Accuracy Produced after PCA (n = 2) on Datasets with Five Classes

Sr. No.	Data Type (GitHub Python File Name)	Algorithm Name	KDD Cup99 Result in %	NSL-KDD Result in %
1.	Numerical (IDS-Bal-PCA)	Logistic Regression	88.19	51.93 (Iteration Reached Limit)
		Decision Tree	99.86	93.38
		Artificial Neural Network	79.13 (Abnormal Termination)	36.38 (Abnormal termination)
		Random Forest	99.87	93.84
		Gradient Boosting	99.37	92.14

TABLE 7.10

Accuracy Produced after PCA (n = 2) on Random Oversampled Data with Five Classes

Sr. No.	Data Type (GitHub Python File Name)	Algorithm Name	KDD Cup99 Result in %	NSL-KDD Result in %
1.	Numerical (IDS-O-PCA)	Logistic Regression	36.03 (Iteration Reached Limit)	21.31 (Iteration Reached Limit)
		Decision Tree	99.37	91.05
		Artificial Neural Network	20.03 (Abnormal Termination)	20.13 (Abnormal Termination)
		Random Forest	99.37	91.06
		Gradient Boosting	90.35	82.20

TABLE 7.11

Accuracy Produced after PCA ($n = 2$) on SMOTE Data with Five Classes

Sr. No.	Data Type (GitHub Python File Name)	Algorithm Name	KDD Cup99 Result in %	NSL-KDD Result in %
1.	Numerical (IDS-OS-PCA)	Logistic Regression	37.38(Iteration limit reached)	20.87 (iteration limit reached)
		Decision Tree	98.6	90.04
		Artificial Neural Network	20.03(Abnormal Termination)	20.13 (abnormal termination)
		Random Forest	98.66	90.29
		Gradient Boosting	90.61	81.44

Finally, Table 7.11 shows the accuracy results on data with five classes and two principal components when the dataset was oversampled using Synthetic Minority Over-sampling Technique (SMOTE).

From Tables 7.8–7.11 it can be observed that the results produced by the identified machine learning techniques after PCA has been carried out, in all the cases are very good results. These results are comparable with the results achieved when no PCA had been done. So, it can be said that here that PCA plays a positive role by reducing the time and computation complexity for the implementation of machine learning techniques without compromising on accuracy.

Another interesting thing to be noted from the Tables 7.8–7.11 is, the results produced by the Decision Tree technique and the Random Forest are very similar, with differences being in fractions. The reason for this similarity in accuracy is that during training the Random Forest technique creates a number of decision trees with different combinations of features but, in this case as the number of features has been reduced to only two there is little space for the technique to create multiple combinations of features. This is the reason why, when the principal components were only two, the results produced by the Decision Tree algorithm and Random Forest techniques are very close to each other. It should also be noted that the differences in the results produced by these two techniques widens with the increase in the number of principal components.

7.5 Conclusion

The high financial impact and sophistication of cyber-attacks mandate that state-of-the-art cyber-attack detection systems are put in place. One of the more effective methods for detection of cyber-attacks is the Intrusion Detection System.

With the sophistication of and evolution in the attacks, even intrusion detection systems are required to evolve and be effective in detecting an attack. Hence, many new technologies have been incorporated into intrusion detection systems in order to make them a potent force in the detection of intrusion even in dynamically changing and challenging scenarios. One of the important technologies incorporated is machine learning. Machine learning techniques have shown great promise in the detection of intrusions based on the anomalies.

In this chapter a detailed analysis of five machine learning techniques: Logistic Regression, Decision Tree, Artificial Neural Network, Random Forest and Gradient Boosting, in detecting intrusions has been carried out. Of these machine learning techniques, Logistic Regression and Decision Tree are legacy techniques whereas, Artificial Neural Network, Random Forest and Gradient Boosting are considered to be state-of-the-art techniques. This evaluation and analysis of the identified machine learning techniques has been carried out on KDDCup99 and NSL-KDD datasets, two of the most popular intrusion detection datasets available.

The performance of a machine learning model depends on a number of factors including: the type of input data, number of classes for classification, the number of training instances for each class and the number of features, among other things. Based on these four parameters a detailed analysis and evaluation has been conducted in this chapter.

From the results obtained, it can be concluded that of the five machine leaning techniques identified, Logistic Regression and Artificial Neural Network techniques were not able to handle the large amount of data, in many cases. This was found to be true especially when the data was oversampled in order to work around the imbalances in the datasets. So, it can be said that for the type of dataset available for intrusion detection system these two machine learning techniques are not very effective.

With the increase in the data, the optimization required for finding the correct results requires a very large number of iterations. With KDDCup99 and NSL-KDD both having very large amounts of data instances, the Logistic Regression technique was required to perform an unusually large number of iterations. This is one of the reasons for the ineffectiveness of Logistic Regression in the present case.

While Logistic Regression required optimization, the Artificial Neural Network required a Back Propagation technique for getting results. Even here, it can be said that the Back Propagation technique couldn't handle large number of dataset instances present for training purposes.

On the other hand, all the three remaining machine learning techniques, Decision Tree, Random Forest and Gradient Boosting produced good results in all the scenarios put forward in this chapter. Decision Tree, Random Forest and Gradient Boosting could not only effectively handle the large amount of data, but also produced very good results.

Another important thing to be noted here is that these three machine learning techniques have produced good results for both types of data, numerical and categorical. One reason for this success is, unlike Logistic Regression and Neural Network, these three techniques are built to handle both numerical and categorical data.

Comparatively, among these three techniques the Random Forest technique performs best. A closer analysis of the results produced by these three techniques proves this point. Even though the results produced by the Random Forest and Decision Tree models are similar, it can be observed that the Random Forest technique consistently outperforms the Decision Tree and Gradient Boosting techniques for both datasets. So, it can be safely concluded that for the datasets used for anomaly-based intrusion detection systems, KDDCup99 and NSL-KDD, Random Forest technique seems to be the best fit.

References

[1] Editors. 2020. Internet Usage Statistics, The Internet Big Picture, World Internet Users and 2020 Population Stats. https://www.internetworldstats.com/stats.htm (accessed Sept 6, 2020).

[2] Ministry of Electronics & IT. Digital Transactions in India. https://pib.gov.in/PressReleaseIframePage.aspx?PRID=1897272#:~:text=BHIM%20UPI%20has%20emerged%20as,lakh%20crore%20in%20January%202023.&text=%23%20Note%3A%20Digital%20payment%20modes%20considered,%2C%20RTGS%2C%20PPI%20and%20others (accessed June 4, 2023)

[3] Federal Bureau of Investigation, Intern Crime Complaint Center. 2020. 2019 Internet Crime Report. https://www.fbi.gov/news/pressrel/press-releases/fbi-releases-the-internet-crime-complaint-center-2019-internet-crime-report (accessed Jan 15, 2020).

[4] Doke, S. 2013. One-Third Internet Users in India Susceptible to Malware Attacks. https://www.firstpost.com/tech/news-analysis/one-third-internet-users-in-india-susceptible-to-malware-attacks-3638661.html (accessed Sept 6, 2020).

[5] Mohammed, S. 2020. India Most Cyber-Attacked Country. https://www.thehindu.com/news/cities/Hyderabad/india-most-cyber-attacked-country/article30678302.ece (accessed Sept 6, 2020).

[6] Sobers, R. 2020. 110 Must-Know Cybersecurity Statistics for 2020. https://www.varonis.com/blog/cybersecurity-statistics/ (accessed Sept 6, 2020).

[7] Milkovich, D. 2020. 15 Alarming Cyber Security Facts and Stats. https://www.cybintsolutions.com/cyber-security-facts-stats/ (accessed Sept 6, 2020).

[8] Khraisat, A., I. Gondal, P. Vamplew, and J. Kamruzzaman. 2019. Survey of Intrusion Detection Systems: Techniques, Datasets and Challenges. *Cybersecurity* 2:20, Springer Open Access, 1–22.

[9] Ioulianou, P. P., V. G. Vassilakis, I. D. Moscholios, and M. D. Logothetis. 2018. A Signature-Based Intrusion Detection System for the Internet of Things. *International Conference on Information and Communication Technology Forum (ICTF).*

[10] Almutairi, A. H. and N. T. Abdelmajeed. 2017. Innovative Signature Based Intrusion Detection System: Parallel Processing and Minimized Database. *IEEE International Conference on the Frontiers and Advances in Data Science (FADS),* pp. 114–119.

[11] Bolzoni, D. and S. Etalle. 2008. Approaches in Anomaly-Based Intrusion Detection Systems. *Intrusion Detection Systems,* eds. R. Di Pietro and L. V. Mancini. Springer Science Business Media, pp. 1–15.

[12] Kumar, S. A. P., A. Kumar, and S. Srinivasan. 2007. Statistical Based Intrusion Detection Framework Using Six Sigma Technique. *IJCSNS International Journal of Computer Science and Network Security* 7(10), 333–342.

[13] Flior, E., T. Anaya, C. Moody, M. Beheshti, J. Han, and K. Kowalski. 2010. A Knowledge-Based System Implementation of Intrusion Detection Rules. *IEEE Seventh International Conference on Information Technology: New Generations.*

[14] Yu, J., P. Tian, H. Feng, and Y. Xiao. 2018. Research and Design of Subway BAS Intrusion Detection Expert System. *IEEE 3rd Advanced Information Technology, Electronic and Automation Control Conference (IAEAC).*

[15] Smola, A. and S.V.N. Vishwanathan. 2008. *Introduction to Machine Learning.* Cambridge University Press.

[16] Goodfellow, I., Y. Bengio, and A. Courville. 2015. *Deep Learning.* MIT Press.

[17] Halimaa, A. and K. Sundarakantham. 2019. Machine Learning Based Intrusion Detection System. *IEEE Third International Conference on Trends in Electronics and Informatics (ICOEI 2019),* pp. 916–920.

[18] Gautam, R. K. S. and A. Doegar. 2018. An Ensemble Approach for Intrusion Detection System Using Machine Learning Algorithms. *IEEE 8th International Conference on Cloud Computing, Data Science & Engineering,* pp. 61–64.

[19] Park, K., Y. Song, and Y. Cheong. 2018. Classification of Attack Types for Intrusion Detection Systems using a Machine Learning Algorithm. *IEEE Fourth International Conference on Big Data Computing Service and Applications,* pp. 282–286.

[20] Yihunie, F., E. Abdelfattah, and A. Regmi. 2018. Applying Machine Learning to Anomaly-Based Intrusion Detection Systems. *IEEE Long Island Systems, Applications and Technology Conference (LISAT).*

[21] Ahmad, I., M. Basheri, M. J. Iqbal, and A. Raheem. 2018. Performance Comparison of Support Vector Machine, Random Forest, and Extreme Learning Machine for Intrusion Detection. *IEEE Access,* pp. 1–7.

[22] Zhanga, H., K. Linb, W. Chenc, and L. Genyuan. 2019. Using Machine Learning Techniques to Improve Intrusion Detection Accuracy. *IEEE 2nd International Conference on Knowledge Innovation and Invention,* pp. 308–310.

[23] Liang, D., Q. Liu, B. Zhao, Z. Zhu, and D. Liu. 2018. Clustering-SVM Ensemble Method for Intrusion Detection System. *IEEE 8th International Symposium on Next-Generation Electronics.*

[24] KDD Cup 1999 Dataset. http://kdd.ics.uci.edu/databases/kddcup99/kddcup99.html (accessed Nov 18, 2020).

[25] NSL-KDD Dataset. https://www.unb.ca/cic/datasets/nsl.html (accessed Nov 18, 2020).

[26] Tavallaee, M., E. Bagheri, W. Lu, and A. A. Ghorbani. 2009. A Detailed Analysis of the KDD CUP 99 Data Set. *IEEE Symposium on Computational Intelligence for Security and Defense Applications (CISDA).*

[27] Nagarhalli, T. P., A. Save, and N. Shekokar. 2021. Fundamental Models in Machine Learning and Deep Learning. In *Design of Intelligent Applications using Machine Learning and Deep Learning Techniques,* eds. R. S. Mangrulkar, A. Michalas, N. Shekokar, M. Narvekar, and P. V. Chavan. Chapman and Hall/ CRC.

[28] Codes for Learning Techniques for IDS. https://github.com/drtpn/Learning-Techniques-for-IDS (accessed Feb 20, 2021).

[29] Brownlee, J. 2020. Random Oversampling and Undersampling for Imbalanced Classification. https://machinelearningmastery.com/random-oversampling-and-undersampling-for-imbalanced-classification/ (accessed Nov 19, 2020).

[30] Chawla, N. V., K. W. Bowyer, L. O. Hall, and W. P. Kegelmeyer. 2002. SMOTE: Synthetic Minority Over-Sampling Technique. *Journal of Artificial Intelligence Research* 16, AI Access Foundation and Morgan Kaufmann Publishers. https://arxiv.org/pdf/1106.1813.pdf (accessed Nov 19, 2020).

[31] Dataman. 2020. Using Over-Sampling Techniques for Extremely Imbalanced Data. https://towardsdatascience.com/sampling-techniques-for-extremely-imbalanced-data-part-ii-over-sampling-d61b43bc4879 (accessed Nov 19, 2020).

[32] Paritosh Kumar, P. 2020. Computational Complexity of ML Models. https://medium.com/@paritoshkumar_5426/time-complexity-of-ml-models-4ec39fad2770 (accessed Nov 20, 2020).

[33] Jolliffe, I. T. 2002. *Principal Component Analysis.* Second edition, Springer.

Section III

Defending Against Cyber Attack Using Deep Learning

8

Deep Neural Networks for Cybersecurity

Bhavi Dave and Aruna Gawade

Dwarkadas J. Sanghvi College of Engineering, Mumbai, India

CONTENTS

8.1 Introduction

Neural networks, rooted in the Universal Approximation Theorem are transforming disciplines like computer vision and natural language processing providing cutting edge products for advertisement, recommendation systems, voice assistants and so on. Since deep learning approaches are found to be state-of-the-art in several domains, an interdisciplinary approach within the cyber security domain is likely to produce great outcomes in protecting

DOI: 10.1201/9781003408307-11

applications, systems and networks. The need for the enforcement of a security methodology is self-evident, with deep neural networks being an ideal candidate modeling non-linear, complex relationships without imposing restrictions on input variables. The future of machine learning in general [1, 2] and deep neural networks in particular [3] looks to be very promising for the domain of cybersecurity. This chapter aims to explore the same.

8.2 Pitfalls in Traditional Cyber Security

The most imminent threats in this age of the internet and data are the rightful concerns of privacy and security. This section details the common cyber security vulnerabilities and attack strategies faced by modern technology.

8.2.1 Denial-of-Service (DoS) Attacks

Denial-of-Service (DoS) attacks render online resources unavailable by reducing, restricting or preventing access to the intended users. Figure 8.1 shows the anatomy of a DoS attack. The salient goal is to curb the legitimate access to the system—often by means of flooding the victim's system with fraudulent service requests or traffic to overload and exhaust server resources (e.g., RAM, CPU etc.)—which in turn compromises the speed of service, network performance and, in dire cases, leads to the unavailability of the system. DoS can present more opportunities for attacks as messages have to be retransmitted to access the resources of an overloaded system, reducing anonymity and security [4]. While DoS emerges from a single source, Distributed Denial-of-Service attacks (DDoS) assault targets the whole network infrastructure, saturating it with fraudulent traffic. The two types of attack share the common objective of overwhelming the system resources and potentially exploiting its vulnerabilities [5]. Although academia is working toward detecting these attacks [6, 7], when the impact of DDoS attacks on cloud computing is explored, it is found that they often remain undetected even when current security provisions are in place [8].

8.2.2 Social Engineering

Social engineering aims to deceive users by duplicitously claiming to be employees, vendors or support personnel exploiting a naïve user's natural inclination of trust to compromise data security. Traditional protections from malware and viruses will not protect users from the human-based social engineering attack, making its prevention especially difficult. Mitigating imminent threats of social engineering are especially difficult since the relationship between human interactions and security breaches pertaining to social engineering has not been adequately explored beyond anecdotal evidence [9].

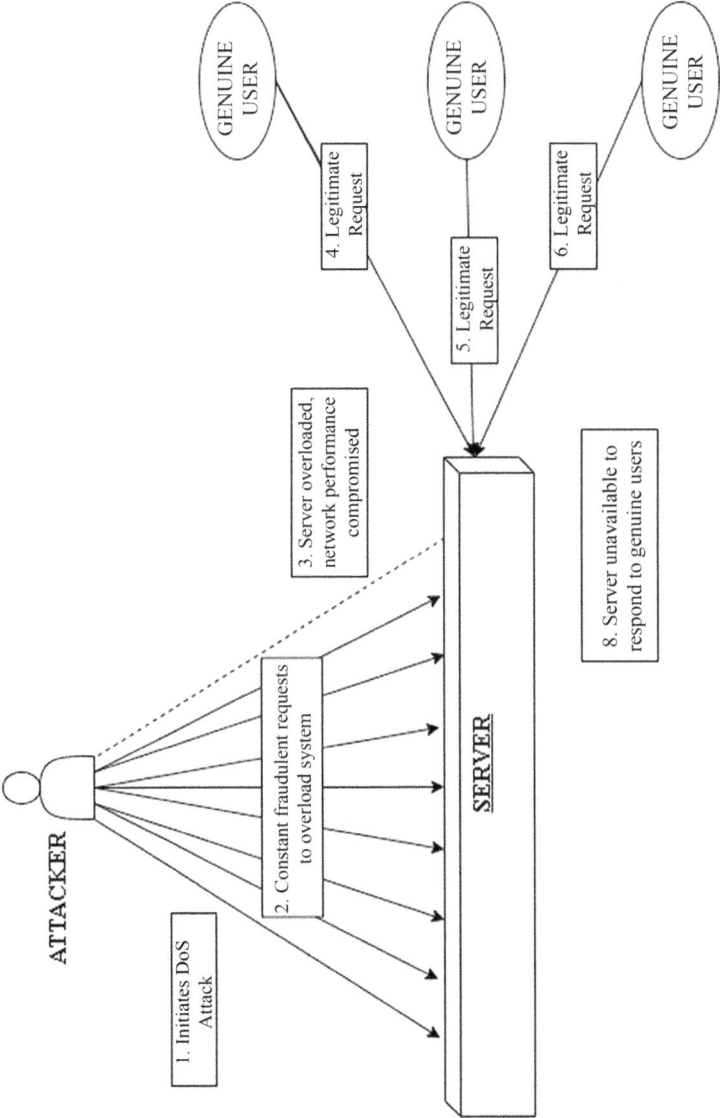

FIGURE 8.1
Denial of service attack.

8.2.3 Phishing

Phishing is an aspect of social engineering where attackers target the human vulnerabilities of end users, often by distributing malicious links or attachments that give the attacker access to personal information, login credentials or even finances when clicked or downloaded. While the average naïve user is especially susceptible to these attacks, even sophisticated users can be tricked with discomfiting ease. Figure 8.2 demonstrates the operation of a phishing site. Empirical experiments depicted that 23% participants did not take even simple security indications in a website into account leading to incorrect choices 40% of the time [10]. Although efforts have been made in detecting phishing sites [11], they are online only for a short period of time, and the fraudster is well on the way to another scam before alarms are raised [12]. This allows them to create a cycle of significant damage without getting caught. Albeit traditional security layers like two-factor authentication, antivirus software, and firewalls do provide protection, they are all fallible and vulnerable to one wrong click.

8.2.4 Malware

Malware, short for 'malicious software', deliberately implements the harmful intent of an attacker [13]. They are seen in different forms like viruses, trojan horses, rootkits or backdoors including adware, spyware and ransomware.

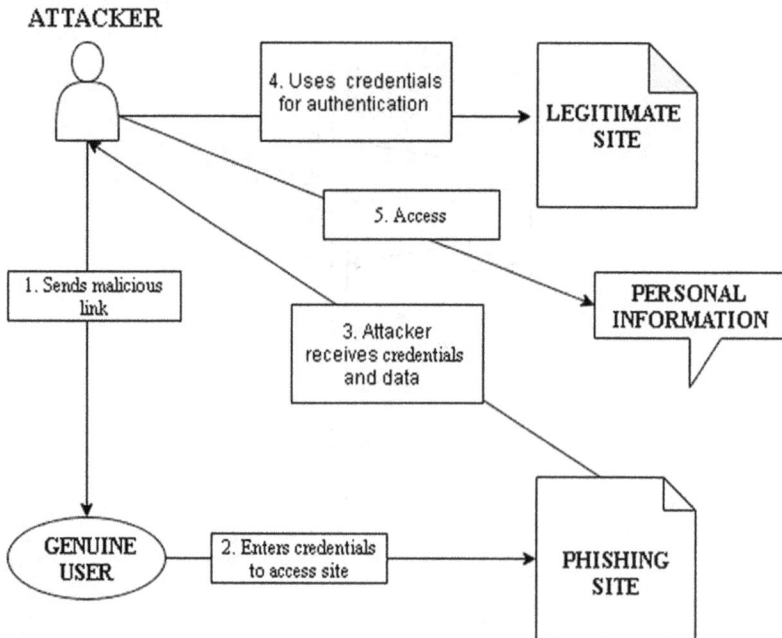

FIGURE 8.2
Attack with a phishing site.

ATTACKER

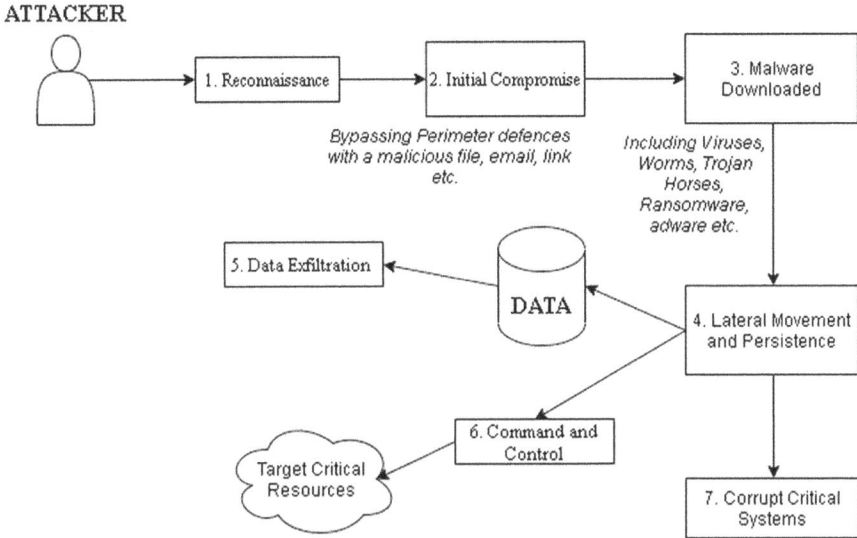

FIGURE 8.3
Malware compromising a system.

Malware is, therefore, an umbrella term used for a variety of hostile or malicious programs. Malware is continually evolving to become more pernicious and sophisticated, not just in sheer volume but also in efficacy by designing varied obfuscation techniques to go undetected. Figure 8.3 displays how malware can compromise the system. A survey conducted by Symantec in February 2019 found that 47% of the organizations had experienced malware security incidents/network breaches in the previous year [14]. McAfee records over 100,000 new malware samples every single day which amounts to a new threat every second. The massive diversity and volume of malware variants renders traditional signature-based techniques ineffective.

8.2.5 Data Breach

Data breach is the disclosure of confidential information to unauthorized parties. The modern technological landscape undeniably relies on the data, making it an essential commodity to enterprises, organizations and institutions alike. Although promising attempts have been made by research for protecting data [15, 16], data breaches have become a pressing concern for enterprises because of the exponential increase in consumer data and the unavoidable digitization of sensitive records and transactions. Figure 8.4 demonstrates a data breach. According to IBM's 2016 Cost of Data Breach Study, the average consolidated cost of a data breach is $4 million [17]. Despite a plethora of efforts, the mitigation of data breaches continues to be a research

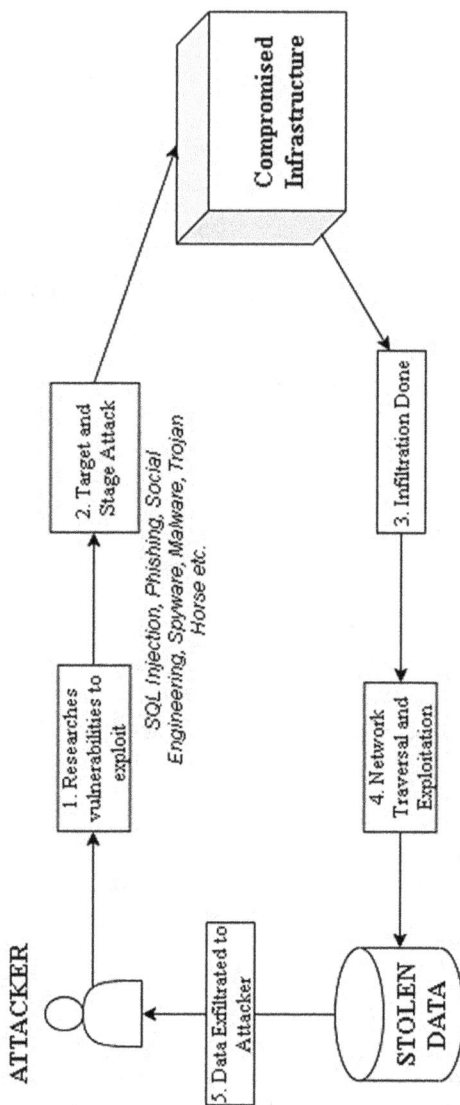

FIGURE 8.4
Data breach attack.

problem since data breaches occur in varied ways [18]. Compromised cyber security infrastructure can inflict damage in the form of data breaches, and consequently, unreliable operations. Secure provisioning of devices, secured connectivity between devices and secure data storage is a non-negotiable aspect of the modern technological landscape.

This section therefore highlights the shortcomings of traditional cyber security necessitating intervention and improvement in the domain.

8.3 Proposed Deep Learning Architectures and Methodologies

Neural networks are deep learning algorithms that learn in a hierarchical manner: each layer continues to gain the ability to learn increasingly abstract features. This section describes and analyses different neural network models, algorithms and methodologies along with their variants.

8.3.1 Convolutional Neural Networks

A convolutional neural network is a subtype of deep neural networks which utilizes alternating layers of convolutions and pooling for every layer, alongside filter banks made up of kernels that are trainable. CNNs generally employ the Rectified Linear Unit activation function that converts negative values to 0, extracting only the prominent features [19]. The max pooling layer helps reduce complexity while allowing prominent features to pass through to upper layers of the network [20]. This process is repeated to gain relevant, abstract and location invariant features that are accurate depictions for recognition problems. CNNs greatly reduce the number of parameters when compared to traditional feedforward networks because of their unique design consisting of weight sharing and pooling. The architecture is especially good at learning feature hierarchies from data [21]. The convolutional neural networks have been improved and altered to generate several variants like RCNN [19], and Fast-RCNN [20].

8.3.2 Recurrent Neural Networks

Recurrent neural networks are useful for remembering information through time, exhibiting temporal dynamic behavior [22]. This class of artificial neural networks can deal with variable length input using its internal state. These internal states can have feedback connections or gated memory, leading to the creation of Long Short Term Memory networks (LSTMs) or Gated

Recurrent Units (GRUs). Connections in an RNN move beyond the traditional feed-forward structure, connecting the output of a neuron back to itself. The architecture can therefore operate over any sequence of vectors, often used with both sequences in the input and the output [23]. Training RNNs can be challenging because the gradients can vanish or explode with reactive ease, failing to achieve the intended purpose of the training algorithm [24].

8.3.3 Generative Adversarial Networks

Generative Adversarial Networks (GANs) learn by the architecture of adversarial training or 'learning by comparison'. The architecture was developed by Goodfellow et al. [25]. There are two parts that work in conjunction in a GAN: the generator and the discriminator. At the end of training, the generator can produce data that so closely resembles the original sample that the generated data is by all practical means indistinguishable from the original source. Furthermore, GANs are widely known for their applicability, which includes, but is not limited to, optical flow estimation, caption generation and image enhancement. A deep convolutional generative adversarial network (DCGAN) [26] is an open-source, pretrained GAN for image generation released by Facebook.

While novel designs strive to surpass the limits of established model architectures, it is noteworthy that these foundational deep learning architectures, and their various iterations, persist as a reference framework, elucidating the rationale behind certain solutions and serving as benchmarks for mathematically derived expectations.

8.4 Deep Learning Applications in Cyber Security

Here is an academic overview of solutions proposed for some of the aforementioned threats, including Intrusion Detection Systems (IDS/IPS), Network Traffic Analytics, Social Engineering, and Malware Detection. These examples illustrate how variants of aforementioned DL techniques are used in cybersecurity.

8.4.1 Intrusion Detection Systems (IDS/IPS) with Network Traffic Analytics

Intrusion Detection and Prevention Systems detect and prevent hackers from gaining unauthorized access to a system while providing reporting and notification functionality. Deep learning approaches have seen a lot of success in analyzing network traffic with better accuracy and reducing the number of false alerts. Deep learning and artificial neural networks are showing promising results in analyzing network requests and HTTPS network traffic to look for malicious activities. Convolutional neural networks have been very successful at the task by virtue of their ability to extract features while simultaneously avoiding overfitting due to weight sharing and pooling. Dilated convolutional autoencoders (DCAEs) proposed by Yu et al. [27] can automatically learn apposite features from raw network traffic data that is large scale and multi-varied to enable network intrusion detection models.

8.4.2 Social Engineering Detection

Social engineering attacks are easy to create and scale since they exploit the weakest cyber security link—human intervention. Given the especially pernicious nature of these attacks, researchers have been continually attempting to create robust systems that can protect users. Deep learning in general and Natural Language Processing (NLP) in particular has had considerable success in detecting spam and other forms of social engineering. Practical, easy-to-use tools are continually developed to protect naïve users from falling prey to social engineering schemes. Researchers built a Chrome extension that analyzed and classified URLs to isolate the malicious ones [28]. While malicious links, spam emails and phishing websites are more commonly known, one must not forget the traditional forms of communication. Deep learning algorithms produce staggering results even in the realm of speech. One such example is the Social Engineering Defence Architecture (SEDA) [29] that enables computer systems to analyze phone conversations in real time to determine if the caller is deceiving the receiver. The work generates real-time signatures while also providing impressive results.

8.4.3 Malware Detection

Deep learning algorithms are capable of detecting more advanced forms of malware as they are not reliant on feature engineering or knowing the signatures of previously identified malware. A comparative study explored the efficacy of classical MLAs and deep learning architectures for Windows malware detection and found that the latter outperformed the former in all types of experiments and could detect the malware executable as malicious or benign within five seconds of execution [30]. The research is also continuing to explore avenues to improve established neural architectures for producing better accuracies. Novel approaches are continually proposed that use innovative techniques alongside deep learning algorithms to create more robust methodologies. Ding et al. [31] applied Deep Belief Nets (DBNs) to detect malware using a dataset created from PE files from the internet. By virtue of learning multiple layers of features with unsupervised learning before connecting to a feed-forward neural network, the architecture was less prone to overfitting and produced an accuracy of 96.1%. As discussed in the previous section, the spread of malware is an imminent threat to the adoption of cloud computing in the present technological landscape.

This section demonstrably illustrates that deep learning methodologies, when applied to imminent problems in the cyber security landscape, produce models that are capable of preventing attacks and detecting breaches.

8.5 Drawbacks and Future Scope

This section provides the challenges faced by the existing academic research and open issues that need to be resolved before the former methodologies are implemented at scale. A paradoxical fact is that deep learning can be adopted by both sides of the cyber battle, and the very concept that is improving modern cybersecurity capabilities is also responsible for the variations and scale of automated attacks. Deep learning and its corresponding approaches can be used to bypass deep-learning based detection systems. The innate lack of interpretability in deep-learning based architectures proposed a non-trivial challenge in its adoption. Since these methods are black boxes [32], errors are difficult to diagnose and attribute to a confirmed cause. Another hindrance in the adoption is the fact that most prevalent approaches focus on solving only one threat, instead of providing a holistic solution for the cybersecurity landscape. This lack of generalization across a broader range of attack vectors [33] would likely not scale to adoption in the industry because of performance constraints.

8.6 Conclusion

Attacks against modern day cyber systems continue to emerge at a rate that vastly outpaces the ability of traditional cybersecurity algorithms. There have also been simultaneous advances made in the domain of deep learning, providing an opportunity to create approaches, prototypes, frameworks and architectures that address the requirements for robust protection and cybersecurity. Several researchers in academia rose to the occasion, providing varied techniques and approaches that address the gap in cybersecurity. Given the especially sophisticated and pernicious nature of modern-day cyber-attacks, the next wave of innovation in security might well be a deep learning system that is inherently more scalable, reliable and accurate. Moving past the current manual process of validating security risks, approaches based on deep learning are more likely to detect imminent threat using the data distributed across the environment including multiple platforms, servers, smart devices, and so on, that encompass the modern-day cyberspace.

References

[1] Srivastava, Kriti, and Narendra Shekokar. "Machine learning based risk-adaptive access control system to identify genuineness of the requester." In *Modern Approaches in Machine Learning and Cognitive Science: A Walkthrough*, pp. 129–143. Springer, Cham, 2020.

[2] Srivastava, Kriti, and Narendra Shekokar. "Design of machine learning and rule based access control system with respect to adaptability and genuineness of the requester." *EAI Endorsed Transactions on Pervasive Health and Technology* 6, no. 24 (2020): e1.

[3] Berman, Daniel S., Anna L. Buczak, Jeffrey S. Chavis, and Cherita L. Corbett. "A survey of deep learning methods for cyber security." *Information* 10, no. 4 (2019): 122.

[4] Borisov, Nikita, George Danezis, Prateek Mittal, and Parisa Tabriz. "Denial of service or denial of security?" In *Proceedings of the 14th ACM Conference on Computer and Communications Security*, pp. 92–102. 2007.

[5] Antunes, João, Nuno Ferreira Neves, and Paulo Jorge Veríssimo. "Detection and prediction of resource-exhaustion vulnerabilities." In *2008 19th International Symposium on Software Reliability Engineering (ISSRE)*, pp. 87–96. IEEE, 2008.

[6] Potluri, Sirisha, Monika Mangla, Suneeta Satpathy, and Sachi Nandan Mohanty. "Detection and prevention mechanisms for DDoS attack in cloud computing environment." In *2020 11th International Conference on Computing, Communication and Networking Technologies (ICCCNT)*, pp. 1–6. IEEE, 2020.

[7] Shekokar, Narendra, Kunjita Sampat, Chandni Chandawalla, and Jahnavi Shah. "Implementation of fuzzy keyword search over encrypted data in cloud computing." *Procedia Computer Science* 45 (2015): 499–505.

[8] Riquet, Damien, Gilles Grimaud, and Michaël Hauspie. "Large-scale coordinated attacks: Impact on the cloud security." In *2012 Sixth International Conference on Innovative Mobile and Internet Services in Ubiquitous Computing*, pp. 558–563. IEEE, 2012.

[9] Workman, Michael. "Gaining access with social engineering: An empirical study of the threat." *Information Systems Security* 16, no. 6 (2007): 315–331.

[10] Dhamija, Rachna, J. Doug Tygar, and Marti Hearst. "Why phishing works." In *Proceedings of the SIGCHI Conference on Human Factors in Computing Systems*, pp. 581–590. 2006.

[11] Shekokar, Narendra M., Chaitali Shah, Mrunal Mahajan, and Shruti Rachh. "An ideal approach for detection and prevention of phishing attacks." *Procedia Computer Science* 49 (2015): 82–91.

[12] Mohammad, Rami M., Fadi Thabtah, and Lee McCluskey. "Phishing websites features." School of Computing and Engineering, University of Huddersfield (2015).

[13] Bayer, Ulrich, Andreas Moser, Christopher Kruegel, and Engin Kirda. "Dynamic analysis of malicious code." *Journal in Computer Virology* 2, no. 1 (2006): 67–77.

[14] Talukder, Sajedul, and Zahidur Talukder. "A survey on malware detection and analysis tools." *International Journal of Network Security & Its Applications* 12, no. 2 (2020. Available at SSRN: https://ssrn.com/abstract=3901568

[15] Shekokar, Narendra, and Vijay Maruti Shelake. "An enhanced approach for privacy preserving record linkage during data integration." In *2020 6th International Conference on Information Management (ICIM)*, pp. 152–156. IEEE, 2020.

[16] Mangla, Monika, Rakhi Akhare, and Smita Ambarkar. "Context-aware automation based energy conservation techniques for IoT ecosystem." In *Energy Conservation for IoT Devices*, pp. 129–153. Springer, Singapore, 2019.

[17] Ponemon Institute. "2016 Cost of data breach study: Global analysis." (2016).

[18] Cheng, Long, Fang Liu, and Danfeng Yao. "Enterprise data breach: Causes, challenges, prevention, and future directions." *Wiley Interdisciplinary Reviews: Data Mining and Knowledge Discovery* 7, no. 5 (2017): e1211.

[19] O'Shea, Keiron, and Ryan Nash. "An introduction to convolutional neural networks." *arXiv preprint arXiv:1511.08458* (2015).

[20] Wang, Bin, Yu Liu, WenHua Xiao, Zhihui Xiong, and Maojun Zhang. "Positive and negative max pooling for image classification." In *2013 IEEE International Conference on Consumer Electronics (ICCE)*, pp. 278–279. IEEE, 2013.

[21] Yamashita, Rikiya, Mizuho Nishio, Richard Kinh Gian Do, and Kaori Togashi. "Convolutional neural networks: An overview and application in radiology." *Insights into Imaging* 9, no. 4 (2018): 611–629.

[22] Miljanovic, Milos. "Comparative analysis of recurrent and finite impulse response neural networks in time series prediction." *Indian Journal of Computer Science and Engineering* 3, no. 1 (2012): 180–191.

[23] Dupond, Samuel. "A thorough review on the current advance of neural network structures." *Annual Reviews in Control* 14 (2019): 200–230.

[24] Bengio, Yoshua, Patrice Simard, and Paolo Frasconi. "Learning long-term dependencies with gradient descent is difficult." *IEEE Transactions on Neural Networks* 5, no. 2 (1994): 157–166.

[25] Goodfellow, Ian, Jean Pouget-Abadie, Mehdi Mirza, Bing Xu, David Warde-Farley, Sherjil Ozair, Aaron Courville, and Yoshua Bengio. "Generative adversarial nets." *Advances in Neural Information Processing Aystems* 27 (2014).

[26] Radford, Alec, Luke Metz, and Soumith Chintala. "Unsupervised representation learning with deep convolutional generative adversarial networks." *arXiv preprint arXiv:1511.06434* (2015).

[27] Yu, Yang, Jun Long, and Zhiping Cai. "Network intrusion detection through stacking dilated convolutional autoencoders." *Security and Communication Networks* 2017 (2017).

[28] Shivangi, S., Pratyush Debnath, K. Sajeevan, and D. Annapurna. "Chrome extension for malicious urls detection in social media applications using artificial neural networks and long short term memory networks." In *2018 International Conference on Advances in Computing, Communications and Informatics (ICACCI)*, pp. 1993–1997. IEEE, 2018.

[29] Hoeschele, Michael, and Marcus Rogers. "Detecting social engineering." In *IFIP International Conference on Digital Forensics*, pp. 67–77. Springer, Boston, MA, 2005.

[30] Vinayakumar, R., Mamoun Alazab, K. P. Soman, Prabaharan Poornachandran, and Sitalakshmi Venkatraman. "Robust intelligent malware detection using deep learning." *IEEE Access* 7 (2019): 46717–46738.

[31] Ding, Yuxin, Sheng Chen, and Jun Xu. "Application of deep belief networks for opcode based malware detection." In *2016 International Joint Conference on Neural Networks (IJCNN)*, pp. 3901–3908. IEEE, 2016.

[32] Rudin, Cynthia. "Stop explaining black box machine learning models for high stakes decisions and use interpretable models instead." *Nature Machine Intelligence* 1, no. 5 (2019): 206–215.

[33] Berman, Daniel S., Anna L. Buczak, Jeffrey S. Chavis, and Cherita L. Corbett. "A survey of deep learning methods for cyber security." *Information* 10, no. 4 (2019): 122.

9

Deep Learning in Malware Identification and Classification

Amruta Hingmire, Priyanka Bhatele, Jyoti Mante, Swati Sinha, Swati Jadhav, Pallavi Shimpi and Uma Pujeri

MIT World Peace University, Pune, India

CONTENTS

9.1 Introduction

"Malware" is a combination of two words: "malicious" and "software", and this highlights how it can cause significant security threats in this computing world. Viruses, spyware, worms, trojans, adware and other similar attack mechanisms are all classified as malware. Malware can be defined as instructions that are injected or penetrate into software applications with the intention to attack the security of the system. The cyber world is an important and

DOI: 10.1201/9781003408307-12

crucial part of today's world but there are many security threats associated with it. We become part this cyber world, by opting into services like banking, communication and networking and e-commerce, among others. Malware has been growing rapidly over time, causing extensive financial losses in the computing world. Different solutions are proposed by various anti-malware companies and organizations to secure systems against malware attack. Due to the substantial growth and complexity of malware, its identification and classification present major challenges for the anti-malware community. Antivirus agencies and researchers are continually exploring new effective and efficient solutions to fight against malware. A number of modern anti-virus solutions have been implemented with machine learning (ML) algo-rithms and are progressing toward deep learning (DL) techniques to protect users from malware [1]. The most popular methods used for malware analy-sis are static and dynamic. This chapter highlights basic methods of malware analysis by applying deep learning strategies for malware classification and visualization. It provides a systematic explanation of static and dynamic approaches for malware detection, shows how to convert malware into an image, gives an overview of neural-network based detection approaches and malware classification using Convolutional Neural Networks, and finally summarizes with concluding observations on different methods. The key objective of this chapter is to provide a reference point for new researchers to explore the domain of malware classification.

9.2 Malware and Its Variants

Malicious software or malware is a program or a file which is harmful to the user's computer/data. It is designed intentionally to gain access to the user's resources by altering and stealing sensitive or confidential data such as password or applications without notifying the user. Malware and different threats are especially invented programs used to perform harm-ful activities. An attacker's intention in designing malicious software is to contact computer services, gain access to sensitive and confidential data by restricting access, affect the device performance, and damage applications running on it.

In the past few years, there has been a significant rise in cyber-attacks [1]. Malware researchers and analysts have proposed various security protocols and methods by designing innovative techniques to combat malware and its variants [2]. There have been many improvements and modifications in the design of malware, and some types are listed below:

Viruses: A computer virus is a form of malevolent software or malicious code written to change the way computer operations are performed and is intentionally designed to spread infection from one machine to another, that is, from computer to computer. It is self-replicable.

Worms: A computer worm is a program that is able to replicate itself without human intervention and spreads copies from computer to computer. Worms can get into users' computers through email attachments or instant messages. These cause havoc to computer resources and could provide access to hackers.

Trojans: A trojan is a malicious program that is pretending to be legitimate software. Trojans are distributed as regular software or games that users install on their computers. Trojans can be employed by cyber-attackers to gain access to users' systems. They can be propagated through some e-mail attachment, spam and pop-ups as well. Trojan horses are malevolent on computers and can cause considerable damage. Cyber security experts consider trojans to be the most dangerous type of malware [3, 4].

Spyware: Spyware is undesirable software that penetrates into a user's computer device to pilfer the user's internet usage statistics, patterns and confidential information without the user knowing.

Bots: A bot, also called an internet bot (a shortened form of internet robot), is a software program that operates as an agent on the internet and is intended to perform repetitive tasks. It simulates human activity by automating certain tasks without specific instructions from humans.

Ransomware: Ransomware is a form of malicious software, which restricts access to your own system or device and encrypts user data. A user may log onto his/her system just like any other day, but then realize that the screen is inaccessible with a pop-up message asking for money or cryptocurrencies. The attacker then forces the user to pay in order to get access to their files or data. This new form of cyber blackmail is one of the more dangerous forms of malware nowadays.

Rootkit: A rootkit is a type of software that is malevolent and allows unauthorized access to computer privileges. Different malicious tools that rootkits contain are user credential stealers, password stealers and Distributed Denial-of-Service (DDoS) [5]. After installing the rootkit, the files are executed remotely, and the system configurations of the host machine are altered. Rootkits are unable multiply themselves or replicate.

9.3 Current Malware Statistics

This section deals with statistics for current malware and its variants. The number of malware attacks are increasing each year. The amount of malware is found to be increasing because they are created using various automation tools and reusing code modules. As per the statistics given by AV-Test institute, "Near about 350,000 new malware and potentially unwanted applications are registered everyday" [6]. Figure 9.1 shows the statistics for malware.

As malware threats have affected most of the computing world domain, analyzing it manually is not a feasible task and, thus, the process of malware analysis must be automated.

Methods for detecting malware can be static, dynamic or hybrid. Static feature extraction means extracting features from a static file, while dynamic feature extraction means extracting the features from malware when it is executing. Static features extracted are used further for classification.

Commercial antivirus software uses signature-based detection for malware detection. This is based on using the static approach of feature extraction. Predefined patterns or signatures of specific malware are stored in a database, and suspicious malware attacks are observed and their patterns are checked against the existing store signatures in the database. A malicious attack which is new and one whose signature or pattern is not available in

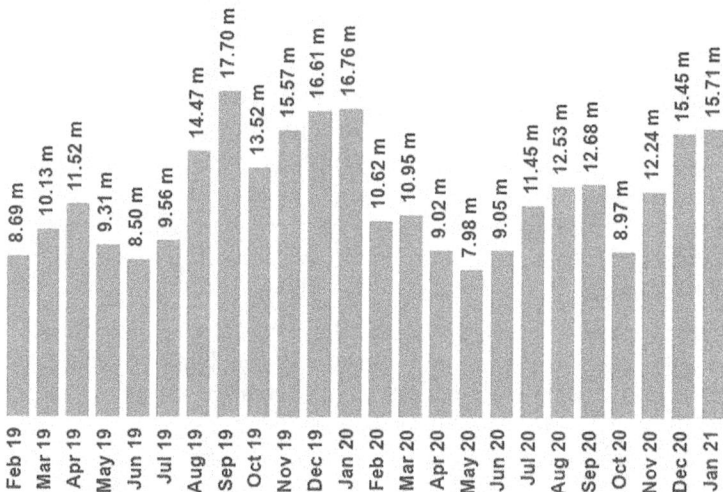

FIGURE 9.1
Malware statistics from Feb 2019 to Jan 2021

(Source: AV-Test).

database will not be detected by this method. Nonetheless, currently most antivirus software uses only the signature-based method.

To detect unknown malware, we need to use the dynamic method for feature extraction. Hence, the latest antivirus software uses dynamic data to detect unknown malware. For the same reason, we propose a deep learning approach for extracting features sequentially from malicious source code and then classifying them. According to recent security trends, malware authors are trying hard to avoid detection by performing some mutations. So, there are more variations of existing malware than new types of malware.

To find an efficient method for classification and analysis of malware, programs should be categorized into different groups as per their respective families. Malware analysis technique is used for detection of malware by labeling a binary executable as benign or malign. It also aims to figure out its category [7]. Most current malicious software analysis methods fail to detect unknown malware. To defeat these challenges, several deep learning models (DL) which are good at analyzing and utilizing harmful codes are receiving the spotlight [7].

9.4 Malware Detection

The pace of growing species of malware programs has increased problems for users day by day. Detection and classification of potential threats have fueled strengthened demand for development. Constant acceleration of malware construction by hackers using approaches like polymorphism has created a difficult task for security vendors. Unidentified malware is mostly used for attacks. For this reason, protecting terminals from infections becomes difficult. Neural networks have served as an acceptable method in much of the research [8, 9].

Malware can be detected at two stages:

1. Detecting the malware files before execution happens in order to prevent them infecting the terminals.
2. Detecting the infected terminals to reduce/cut down the expansion of the infection.

9.4.1 Anomaly-Based Detection

It can be said that this system has a good capability to detect any abnormal activity compared to normal activity, and also to prevent further attacks with the accuracy of detection mentors. Anomaly-based detection is part of data mining which detects data points, events and/or observations that deviate from a dataset's normal behavior. For example, for large data instances, the distance or similarity is compared in every aspect of measurement and then finally compared by checking variation in parameters and detecting how far the normal parameters are different from abnormal parameters.

9.4.2 Signature-Based Detection

Signature-based detection is used to detect software threats like viruses, worms, trojans, malware and so on. Signature-based detection is useful to detect known threats. A signature to an intruder is a similar or common pattern of the user. For example, an attachment in an email with a very interesting topic will be sent to attract a user, or a remote login as an admin user claiming to be admin itself. An alert message to the user will be generated if a difference between both parameters is observed, or otherwise, data in the network will be allowed to flow normally.

The virus scanners of antivirus programs, scan the computer to identify the signature pattern, or the digital footprint of the file, as these are typically unique to specific properties. Antivirus products makes use of databases which have some signatures of known malware samples/files. During scanning, if the antivirus program discovers one of these patterns, then the file can be flagged as being infected. Signature-based methods are reactive in nature.

The procedure of malware identification using the signature-based method is:

1. An unknown type of malware is detected.
2. The signature or digital footprint of the newly identified malware is added into the database.
3. The antivirus program is updated by including the newly detected signature.
4. The updated version of the antivirus product is now able to detect malware during a computer scan.

The drawback of this method is that if the signature of a new malicious attack is not available in the database, then it will not be detected. Signature-based detection can also be implemented for application programming interface (API)-based tracing.

9.5 Machine Learning in Malware Detection

The fundamental purpose of machine learning (ML) is to recognize patterns in a dataset. In recent years, machine learning algorithms have gained more attention in detecting and analyzing numerous types of malware.

In a supervised learning strategy, the data samples are labeled with their class, and classification is done by selecting different algorithms. Nearest-Neighbor (NN) classifiers have also been implemented by researchers for malicious and benign classification. Behavior-based methods make use of API calls, data flow analysis and multiple path execution [10]. These can be monitored by running in sandbox or virtual environments with use of ML algorithms.

With a growing dataset come issues of scalability. To handle these issues, data need to be rescaled by keeping only relevant features. So, various feature selection algorithms have been implemented to reduce the size of the number of instances. The features are obtained by analyzing code or tracing events such as system calls, registry accesses and network traffic. The extracted features from malware images are then classified using different ML classifiers [11].

We can change a malware/kind-hearted document into a grayscale picture utilizing the technique depicted later. At that point we can apply these profound learning procedures on the produced pictures to characterize them as malware or kind. The selection of features, directly affects the accuracy and computation time.

9.5.1 Neural Networks for Malware Detection

A neural network is a computing model which is a replica of a human brain. Biological neural networks consist of a nerve cell, with a nucleus that processes information inside it. The body of the neuron cell is called a soma cell.

Neurons have an extended cable, called an axon, that passes information further and acts as an output. Information is received by neurons by dendrites and these act as an input to the neuron. Axon and dendrites are in turn connected to each other via synapses. On the contrary, in an artificial neural network (ANN), neurons are connected to each other via connection links. The connection link bears a weight. Weights are the relative values that have information about the input signal. Neurons process information based on the weights as they excite or inhibit the signal that is being communicated. Neurons have an internal state which is called the activation signal. The output signal is generated with the combination of the input signal and the activation rule as shown in Figure 9.2.

9.5.1.1 Connections and Weights

The network is made of connections; each connection provides the output of one neuron as an input to the next neuron in the network. Every link is assigned a weight value which represents its relative importance.

Input Weights

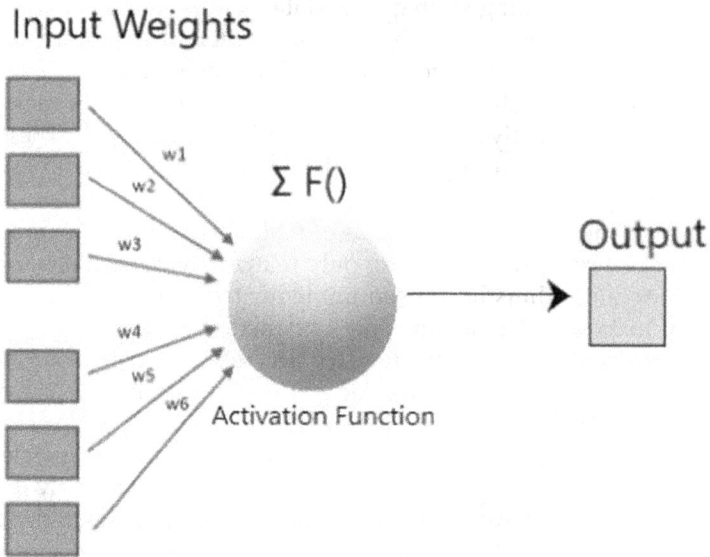

FIGURE 9.2
Artificial neural network.

9.5.1.2 Propagation Function/Activation Function

The propagation function calculates the input to a neuron from the outputs of the neurons before it and the weighted sum of connections associated to it. A bias value supplied to the network, can be added to the result of the propagation. The continuous activation function, f(X) associated with the neurons has a purpose. It makes the system learn new things; as it learns through experiences, connections are modified and new connections are formed.

9.5.1.3 Learning Process

The objective of a classifier is to learn features automatically from input. ANNs are the preferred way to do this, by altering their weight and other parameters in such a way that the resultant output gets much closer to, or the same as, the desired output. Supervised learning is used when we know about the desired output corresponding to each input. In unsupervised learning, the output is not known already.

During the model training stage, input is fed from the dataset one-by-one and the weights are updated accordingly in case of any errors generated. This process learns through mistakes just like the human brain. This process is repeated until the difference between the output generated by our model and actual outputs become very small or negligible. There are several ways of achieving our desired output. One popular way is to start with some random initialization on weight using some distribution.

An input 'x' is supplied to the network and output 'y' is obtained.

$$x(\text{input}) \rightarrow y(\text{output})$$

Subsequently, error is calculated considering how far is y from the actual desired output y_0.

$$E(\text{error}) = y_0 - y$$

Our goal is to minimize this E (error) as far as we can.

Like Convolutional Neural Networks, Deep Neural Network architecture performs feature extraction of a malware, by using autoencoder-based feature learning. Malware classification can also use recurrent neural network-based classifiers.

1. **Convolutional Neural Network (CNN)**

   A Convolutional Neural Network is a popular deep learning algorithm which works on images. CNNs were inspired by the mammals' visual cortex. Neurons in one layer do not connect to all the other neurons in the next layer. Instead, they are connected to only a small portion of it. Output of these networks are reduced by a single vector.

   The main component of CNN architecture is convolution, which combines two functions to produce a third function. Convolution is done with an input vector and 2D array of weights known as a filter or kernel. The feature extraction process or hidden layer is responsible for identifying special kinds of features from the input. In this phase, a complete input image is scanned using a filter to find out whether the expected feature is present anywhere in an image [12].

   The process of applying a filter is just a multiplication (dot product) of filter and filter-sized region of the input image. This is repeated several times at distinct points on the input to cover the entire image and obtain a feature map (in 2D output array) [13].

2. **Recurrent Neural Network (RNN)**

   Recurrent Neural Networks (RNNs) are a type of artificial neural networks which are most suitable for handling sequential data and time series data. They are used in cases where input data possesses some temporal properties such as language translation, transactional data, speech recognition, image captioning, stock market analysis or handwriting recognition [14]. RNNs are networks with memory. Unlike traditional NNs, where output depends upon input, here, output relies on current input information and the historical data collected over the network with the memory units:

$$(\text{INPUT} + \text{PREVIOUS HIDDEN STATE}) \rightarrow \text{HIDDEN STATE} \rightarrow \text{OUTPUT}$$

Input: RNNs do not read all the input data at once. Instead, they take one input at a time and in a sequence.

Hidden State: Each cell of the network is responsible for doing some calculations and hidden state is obtained after combining input and previous hidden state Figure 9.3 shows the basic structure of an RNN, where

$x_1...x_n$: input layer
$h_0....h_n$: hidden layer
$y_1...y_n$: output layer.
w_h, w_x, w_y: network parameter used to improve output of the model.

The basic computation that the RNN's cells does to produce the hidden states and outputs is as follows,

$$ht = (wh * ht - 1 + wx * xt)$$

The result of the above multiplication, that is, *ht* is passed to selected activation function to introduce non-linearity in the result. The resultant *ht'* hidden layer is then multiplied by weight matrix to obtain expected output.

$$yt = wy * ht$$

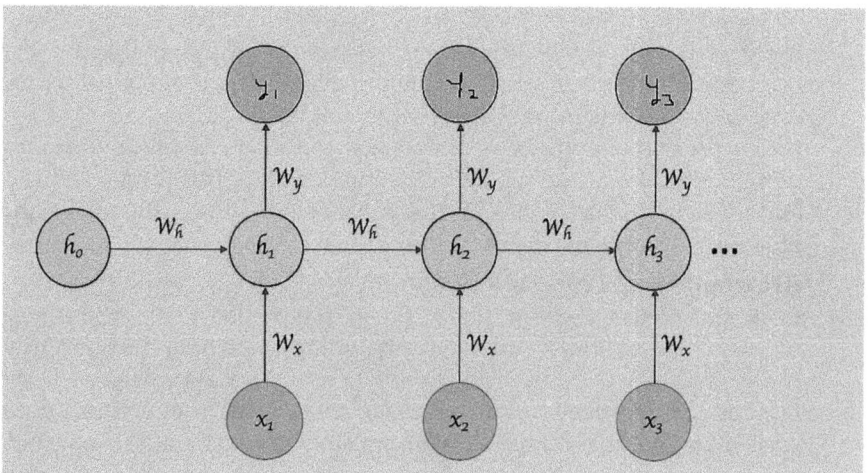

FIGURE 9.3
Recurrent neural network (RNN).

9.5.2 Detection Using Combined Features

This method is best applicable to find whether a portable executable (PE) file is malicious or benign. There are two types of features of a PE file, static and dynamic. Static features are the basic information related to the file such as: byte histogram, byte entropy histogram, section information, imports and exports information, general file and header file information and string extractor. On the other hand, API calls are regarded as dynamic features. Malware detection can be done using a combination of static and dynamic features [15].

The methodology can be split into four stages:

1. The features of the PE files are extracted.
2. Sandbox is used to record the API sequences which are the processes of the RNN.
3. The static and dynamic features are combined and converted into a fixed feature vector. This vector is then transformed to an image.
4. CNN-based models are used to train and classify these images.

Sometimes, different types of malware possess some structural similarities and share some common features. The signature-based identification method is not as efficient in identifying new types of malware, as they do not consider any structural and behavioral similarities. Most recent malware is very similar to already recognized types. The method discussed in this chapter represents a malware binary file as a grey-scale image for analysis and classification [16].

9.6 Malware Visualization and Classification Using Deep Learning

This is the basic outline of the procedure:

- Collection of malware samples and preprocessing
- Visualization of malware as an image
- Training a neural network model for classification
- Testing/validation of a model.

9.6.1 Collection of Malware Samples and Preprocessing

For dataset preparation, malware samples can be collected from various malware repositories such as:

1. MalImg dataset which is uploaded by Nataraj et al. The MalImg dataset was made publicly available by Microsoft for the "Big Data Innovators Gathering Anti-Malware Prediction Challenge"[17].
2. VirusShare: https://virusshare.com/
3. AVCaesar: https://avcaesar.malware.lu/
4. Malshare: https://malshare.com

Currently, there are many repositories which allow open access for researchers. Some portals like, VirusShare, Malshare and VirusTotal, among others, are intended for collection of malware samples using honeypots and also from people around the world, who share their malware files for analysis. These files are also made available to the public. Sometimes, gathered data need to be preprocessed to clean out duplicates by comparing MD5 hash of the malware samples. In Table 9.1, a distribution of different malware samples from MalImg dataset is shown.

TABLE 9.1

MalImg Dataset: Distribution of Samples

Malware Family	No. of Samples
Adialer.C	125
Agent.FYI	116
Allaple.A	2949
Allaple.L	1591
Allueron.gen!J	198
Autorun.K	106
C2Lop.P	146
C2Lop.gen!g	200
Dialplatform.B	177
Dontovo.A	162
Fakerean	381
Instantaccess	431
Lolyda.AA1	213
Lolyda.AA2	184
Lolyda.AA3	123
Lolyda.AT	159
Malex.gen!J	136
Obfuscator.AD	142
Rbot!gen	158
Skintrim.N	80
Swizzor.gen!E	128
Swizzor.gen!I	132
VB.AT	408
Wintrim.BX	97
Yuner.A	800

9.6.2 Visualization of Malware as an Image

To visualize a malware file as a picture, every byte of the file has to be interpreted together as a pixel [18]. The idea of visualizing malicious executable as an image, provides a way to segregate each section of this binary representation. Moreover, malware creators are only changing a small portion of the code from malware files to obtain new variants of it. Thus, if known malware files are mutated to create new binaries, the resulting malware programs would be similar to the originals.

A binary file just displays the file in hex and ASCII formats and does not convey any structural information associated with it. These files are composed of some primitive types. In a binary file the data is represented in a binary form, containing a sequence of zeros and ones.

For visualization of malware, read a given file as a vector of 8 bit unsigned integers at one time and then organize it in the form of two-dimensional arrays. In grayscale images, the value of each pixel indicates the number of bits of data used for representing a pixel. Grey-scale images are usually represented with 8 bits. So, a combination of 8 bit binary digits is the pixel value of a pixel and the range 0–255 indicates that there are a total 256 grayscale levels. Then, convert the malware files into a series of zeros and ones. Upon considering only 8 bits of binary file at a time, an 8 bit vector is produced representing a 2D matrix as a grayscale image, which is nothing but the image representation of a binary file.

The height of each image obtained appears to be varying in size and not the same, as it completely depends upon the size of the converted binary file.

Table 9.2, shows grey-scale image representations of malware belonging to different malware classes.

Here, malware belonging to the same family exhibits some textural similarity compared with another class. Malware from same class is more similar than malware from different classes. Next, convert each file into 256 × 256 grayscale image [13, 19].

9.6.3 Training a Neural Network Model for Classification

A. Feature Selection

Traditional recognition approaches work in two stages: (1) feature extraction phase and classification and (2) techniques for decision making.

Machine learning has proposed several features to analyze the texture of a converted image. Traditional approaches of texture analysis, analyze the texture block for frequency content. Standard methods are adopted to divide these images into rings and wedges. For feature selection, various algorithms such as a homogeneous texture descriptor, texture feature extraction, colour layout descriptor, or Global Image Descriptors (GIST) can be used.

TABLE 9.2

Grey-scale Image Representations of Malwares

Adialer.C			
Allaple.L			
Dontovo.A			
Lolyda.AA3			
Malex.gen!J			
Yuner.A			

To compute texture features, GIST can be a good choice. It divides the image into a four-by-four grid for obtaining the orientation of a histogram. GIST captures textural similarity between images. This feature has recently shown good results in image searches, scene classification and object classification.

B. **Build CNN Network Architecture**
Once the required features are identified, a CNN network can be built. The input to the network is a greyscale image, I w, h, d, where

> w: Width of image
> h: Height of image, and
> d: Depth of image (d = 1)

The sizes of input images are not identical. Thus, rescaling is done before passing the image to the network. The convolution process takes an input signal of size w × h × d, and applies a filter/kernel of size k × k × d, (where k ≤ w, h) giving a single output signal.

The construction of a CNN architecture is as shown in Figure 9.4, with three alternate occurrences of convolution kernels and pooling layers such as Max Pooling, with a fully connected layer (FC) at the end. Convolution layers consist of 50, 70 and 70 filters (kernels) with size of 5 × 5 × 1 for the first, 3 × 3 × 50 for the second, and 3 × 3 × 70 for the last layers, respectively. The model learns spatial hierarchies of features through convolution kernels, ReLU activation and pooling layers.

The activation function is used to learn distinct identification of expected features. The pooling operation is used to reduce the dimensions of the feature map. The Max Pooling layer with filters of size 2 × 2 × 1 with stride-1, reduces number of pixels, thus, reducing the input signal by half [20]. Next, Local Response Normalization (LRN) normalizes the input values in the layer for some neurons [13, 19].

C. **Training the Model**
Neural networks are very popular and successful in feature identification and pattern classification. Before we train the model, the images are resized to a common size of 32 × 32 (i.e., row × column). the dataset consisting of samples from different classes is divided into train and test sets. On performing a split on the dataset, the first part is selected for training and the rest of the samples for testing/validation [21].

Following are the sequence of steps performed on a network to train the model.

Step 1: Initialize weights and other parameters of the network with some arbitrary (random) values.

Step 2: Supply input to the network. The produced output is as a vector of probabilities belonging to each class.

Step 3: Calculate error by looking at some of the probability differences.

Step 4: Use back propagation, to minimize error/cost by adjusting weight and other parameters.

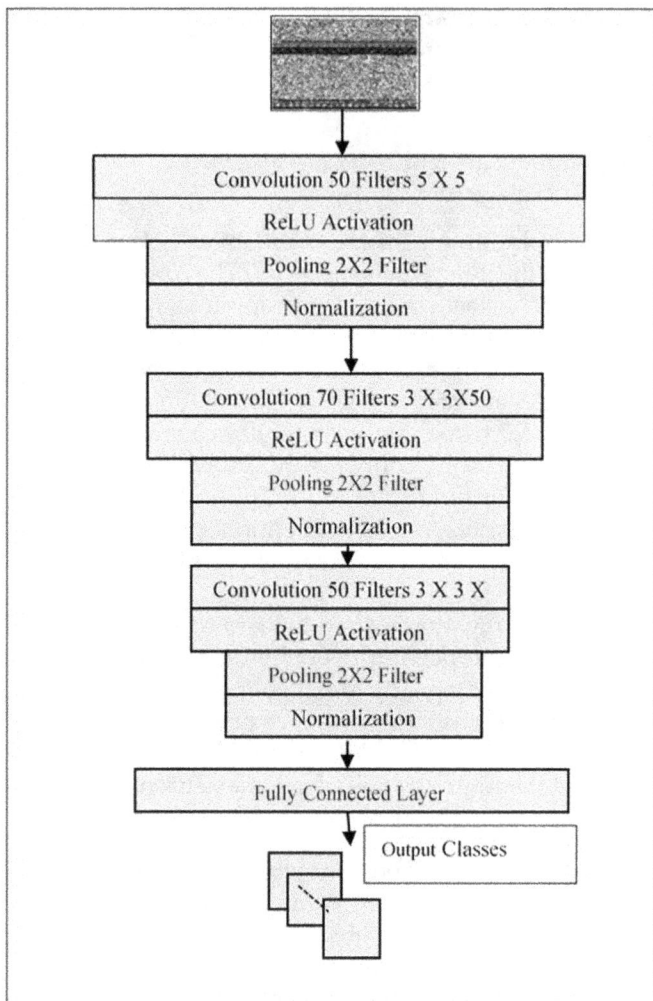

FIGURE 9.4
CNN for classification of malware of grey-scale representation.

Step 5: Modify weights values in proportion to their contribution in generated error.

Step 6: Repeat the process until desired/improved results are obtained.

While training our model, our aim is to always minimize the difference between generated output values and desired output [22].

9.6.4 Testing/Validation of a Model

Use the trained network model to classify any new malware sample. For implementation of the Python system, different python libraries can be utilized. The Python Imaging Library (PIL) of Python, supports image processing functionalities and provides extensive formats. For developing deep learning models, Keras (Neural Network Library) and TensorFlow frameworks work well. CNN classification models give good test accuracy. In the testing and validation phase, an assessment is done of the performance of our deep learning model created for classifying malware. The results showed an impressive accuracy of 96%, indicating that the model is strong and dependable in accurately detecting and categorizing malware. Other variants of neural networks and some hybrid models are also helpful to get improved accuracy over simple CNN.

9.7 Summary

Malware, which is growing in tandem with software and emerging technologies, is continuously adapting new paradigms, allowing a rise in threats of data breaches. Machine learning algorithms are used widely to address the issue of malware detection and its classification. These methods have been proven to be effective in malware detection. To capture the behavior of malware, it is required to execute these malware files. Executing these malware files needs high computation power and time. The static- or behavior-based methods are platform dependent. Hence, separate classifiers are needed for each of the platforms. Deep learning techniques are able to recognize patterns of malware binaries, which can be represented as an image. Malware visualization techniques described in this chapter follow an image-based approach, and are able to achieve significant improvements in performance. With image representation, malware can be analyzed with greater accuracy without executing malware files.

References

[1] Pooja, B., & Kumar, G. 2020. Detection of malware using deep learning techniques. *International Journal of Scientific and Technology Research*, 9, 1688–1691.
[2] Shekokar, N. M., Shah, C., Mahajan, M., & Rachh, S. 2015. An ideal approach for detection and prevention of phishing attacks. *Procedia Computer Science*, 49, 82–91.

[3] Stamp, Mark, Alazab, Mamoun, & Shalaginov, Andrii. 2021. *Malware Analysis Using Artificial Intelligence and Deep Learning*. Springer.

[4] Shekokar, N., Sampat, K., Chandawalla, C., & Shah, J. 2015. Implementation of fuzzy keyword search over encrypted data in cloud computing. *Procedia Computer Science*, 45, 499–505.

[5] Potluri, S., Mangla, M., Satpathy, S., & Mohanty, S. N. 2020, July. Detection and prevention mechanisms for DDoS attack in cloud computing environment. *2020 11th International Conference on Computing, Communication and Networking Technologies (ICCCNT)* (pp. 1–6). IEEE.

[6] AV-TEST The Independent IT-Security Institute. 2021. Malware statistics and trends report by av-test institute. https://www.av-test.org/en/statistics/malware (accessed Feb 2021).

[7] Lee, Y. S. et al. 2018. Trend of malware detection using deep learning. *International Conference on Multimedia and Expo Technology (ICMET)*, ACM.

[8] Shah, M. D., Gala, S. N., & Shekokar, N. M. 2014, April. Lightweight authentication protocol used in wireless sensor network. *2014 International Conference on Circuits, Systems, Communication and Information Technology Applications (CSCITA)* (pp. 138–143). IEEE.

[9] Alazab, M. 2015. Profiling and classifying the behavior of malicious codes. *Journal of Systems and Software*, 100, 91–102.

[10] Du, Y., Wang, J., & Li, Q. 2017. An android malware detection approach using community structures of weighted function call graphs. *IEEE Access*, 5, 17478–17486.

[11] Nagarhalli, Tatwadarshi P., & Shekokar, N. 2021. Fundamental models in machine learning and deep learning. *Design of Intelligent Applications using Machine Learning and Deep Learning Techniques*. Hall/CRC.

[12] Zhou, Huan. 2018. Malware detection with neural network using combined features. *15th International Annual Conference, CNCERT*, Beijing, China.

[13] Gibert, D., Mateu, C., Planes, J. et al. 2019. Using convolutional neural networks for classification of malware represented as images. *Journal of Computer Virology and Hacking Techniques*, 15. 10.1007/s11416-018-0323-0.

[14] Jha, S., Prashar, D., Viet Long, H., & Taniar, D. 2020. Recurrent neural network for detecting malware. *Computers & Security*, 99, 102037.

[15] Tobiyama, S., Yamaguchi, Y., Shimada, H., Ikuse, T., & Yagi, T. 2016. Malware detection with deep neural network using process behavior. *IEEE Computer Software and Applications Conference (COMPSAC)*, Atlanta, GA.

[16] Ajay, S., Kumar, A., & Handa, A. 2020. Malware analysis using image classification techniques. In Shukla, S., and Agrawal, M. (Eds), *Cyber Security in India* (pp. 33–38). IITK Directions, vol 4. Springer, Singapore.

[17] Nataraj, L., Karthikeyan, S., Jacob, G., & Manjunath, B. S. 2011. Malware images: Visualization and automatic classification. *International Symposium on Visualization for Cyber Security*, ACM, New York, NY, USA.

[18] Han, K., Lim, J. H., Im, E. G. 2013. Malware analysis method using visualization of binary files. *Research in Adaptive and Convergent Systems, RACS*.

[19] Ding, Yuxin & Siyi, Zhu. (2019). Malware detection based on deep learning algorithm. *Neural Computing and Applications*, 31. 10.1007/s00521-017-3077-6.

[20] https://machinelearningmastery.com/pooling-layers-for-convolutional-neural-networks/

[21] Vasan, Danish et al. 2020. IMCFN: Image-based malware classification using fine-tuned convolutional neural network architecture. *Computer Networks*, 171, 107138.

[22] Akarsh, S., & Simran, K. 2019. Deep learning framework and visualization for malware classification. *International Conference on Advanced Computing & Communication Systems (ICACCS)*, Coimbatore, India (pp. 1059–1063), IEEE.

Section IV

Defending Against Cyber Attack Using Advance Technology

10

Cyber Threat Mitigation Using Machine Learning, Deep Learning, Artificial Intelligence, and Blockchain

Snehal Paddalwar, Leena Ragha and Vanita Mane
Ramrao Adik Institute of Technology, Mumbai, India

CONTENTS

10.1 Introduction

A cyber threat is an activity that bypasses the security of a system and alters the confidentiality, integrity and availability of the data of a system. Cyber threat activities are carried out in a cyber-threat environment where cyber

DOI: 10.1201/9781003408307-14

actors conduct malicious activities. Cyber threat actors can be individuals or groups who intend to take advantage of system vulnerabilities and, thereby, affect the victim's data and networks. Cyber threat actors are categorized by their intentions of attacking. For example, profit making, ideological violence, geopolitical, and satisfaction are some of the intentions of cyber threat actors in attacking systems and fulfilling their goals. In every aspect of life, whether daily activities such as connecting home devices with the internet, online shopping, banking transactions, and so on, or whether industrial, organizational or institutional level work, each of these aspects is somewhere vulnerable to cyber threats. The severity of cyber threats increases when the system vulnerabilities are known to the attackers. To minimize cyber-attacks or cyber threats, the foremost thing that needs to be done is identifying the system's vulnerabilities. Once the vulnerabilities are known, security patches can be added to make the system more secure and less vulnerable to attacks. There is a non-exhaustive list of cyber threats, as attackers continue to evolve by learning the system's loopholes and thereby expanding the scope of their attacks. Making an organization secure from cyber threats and updated with the latest security-related technologies is a challenging issue. In this chapter, we study the use of technologies such as machine learning (ML), deep learning (DL), artificial intelligence (AI) and blockchain to understand their roles in reducing cyber threats.

The outline of the chapter is as follows. Section 10.2 explores selected research papers to understand the role of these technologies. Section 10.3 gives a basic overview of threats, while various technologies to mitigate these threats are discussed in section 10.4. Section 10.5 discusses a proposed technology-rich cyberinfrastructure to address the cyber threats. This is followed by the conclusion and consideration of future scope.

10.2 Literature Survey

In order to understand the role of ML, DL, AI and blockchain, a literature survey was carried out and it was observed that there is no single solution that can work on all threats. Researchers have used these techniques for identifying various attacks and have claimed good accuracy. In this section, we explore a few research papers to understand the role of these technologies and their impact in addressing the threats.

In paper [1], the authors completed a thorough study of various ML approaches and their utilization to address the threats faced in networked cyber physical systems.

Artificial intelligence and ML show potential in addressing the challenges of cyber security threats and the researchers claim that interaction between

ML and cyber security serve as a defense strategy and not as an attack strategy.

In paper [2], in order to protect data, the researchers used ML and DL technologies by integrating them with cyber-attacks to provide cyber security.

The researchers worked on various cyber threats like breaking captchas, phishing and spamming, and they concluded that ML performs best in handling various cyber-attacks. They also claimed that a system can also be trained to build immunity for various kinds of attacks and the system can also be made to learn from new attacks and protect them from future attacks.

The authors in paper [3] claimed that the methods could be integrated in cyber-detection systems with the goal of supporting, or even replacing, the first level of security analysis. Their conclusions are based on experiments performed on real enterprise systems and network traffic to handle threats such as intrusion, malware and spam.

The authors in this paper propose that DL is still at an early stage and no final conclusions can be drawn, suggesting that further work using adversarial learning may help to improve recognition. There is a risk in using automation to support the security operator accuracy as overestimation of ML capabilities can facilitate skilled attackers to steal data or sabotage the enterprise.

Authors in paper [4] worked on phishing attacks on emails using Support Vector Machine (SVM), AML and DL methods. They observed that DL performed slightly better than ML and also concluded that context of the email content is important in identifying phishing attacks.

In paper [5], the authors proposed a ML model for cyber security in the Internet of Things (IoT). They used and evaluated this model using four ML model with ML algorithms and found that the hybrid Convolution Neural Network (CNN) + Long Short Term Memory (LSTM) performed better with a higher accuracy level (97.16%) than other ML and DL models. Their paper also identified various challenges for usage of ML algorithms for cyber security in the IoT.

In paper [6], the authors introduced an approach that uses ML and visual representation for identifying malicious traffic in a network. After comparing with multiple neural networks the authors have observed that a Residual Neural Network (RNN) with 50 layers is promising in the identification of malware network traffic with 94.50% accuracy.

The authors in the paper say that in future, to improve the protection and mitigation processes in this Intrusion Detection System (IDS), they plan to implement the ML module.

The authors in paper [7] discussed the possibilities of blockchain technology in boosting cyber security. They claimed that it is impossible to break codes and keys in blockchain technologies are they combine anonymous users, computers and dates. They say that blockchain technologies in businesses can be used to authenticate users without providing special information. Also, they state that blockchain is promising in protecting information and operations and it also takes responsibility for protection against strong cyber-attacks.

The authors in paper [8] stated that ML based analytics is an excellent tool to minimize false positive security alerts. They claimed that ML is most suitable for analyzing huge volumes of security events. In addition, they presented various examples of using ML for enhancing cyber security monitoring.

It is observed that the researchers used ML and DL techniques for the same task and compared their accuracies.

In paper [5], the authors tried to use ML and DL combined approaches and they showed that the combined approach had the better performance. As most of these experiments were on the standard latest datasets [9, 10], there was no discussion of how the information could be accumulated in real time in the network. Also, it is observed that there is a need for securing data that may be tampered with by the attackers to breach security and prevent cyber security solutions from catching them. Based on this analysis, it is concluded that one technology alone cannot address the cyber-threat problem. There is a need to use these technologies together by exploiting their strengths to catch the threats. Hence, it is important to understand the threats and the technologies.

10.3 Cyber Threats

In order to mitigate cyber threat, it is necessary to gain awareness of all the cyber-threat related terminologies. What exactly does "cyber threat" mean? Who are cyber-threat actors? Where do cyber-threat actors perform their activities? What might be the strategies of cyber-threat actors? Where might cyber threats occur? What are the sources of cyber threats? What types of cyber threats are commonly performed? and so forth, are some of the questions that need to be answered before planning a cyber-threat mitigation strategy. Following are some answers to all the above questions.

10.3.1 What Is a Cyber Threat?

With the increasing using of the internet, everything has gone online, whether it is a small private conversation or a big business deal transaction. All this data faces the danger of being damaged or stolen by someone else, in order to fulfil their needs or to hinder the progress of an organization. The reasons for stealing or damaging data are numerous.

A cyber threat is an act of finding a way to steal data or damage data. This malicious act can be performed by inserting a computer virus, breaching the data, or carrying out a phishing attack, or Denial of Service (DoS) attack and so forth. A cyber threat can come from people who are working in the same organization or from unknown people in unknown remote locations.

A cyber-attack performed by cyber-threat actors can cause huge disruption to day-to-day activities. It can cause entire data loss, breaching of confidential security, failure of medical equipment, damage to valuable data, paralysis of the network and contamination of data. Cyber-attacks can result in huge financial issues, if not prevented ahead of time.

10.3.2 Cyber-Threat Actors

An enemy of cyber security can be called a threat actor. The threat actor can be an individual or a group of people. These actors are not always outsiders; they could be a part of the same organization which they aim to attack. Such cyber-threat actors always hide their identity and work silently. Some motives of such actors could be to destroy data, hinder the company's progress, or gain money, among others.

Cyber-threat actors always have a goal. Once their goal is defined, they look for a target to achieve their goal. Once the target is defined and set, cyber-threat actors engage themselves in finding a weak area of the security system of an organization. Cyber-threat actors look for the thinnest place where the security can be easily broken. This might also include gaining the trust of the organization or its employees, and later acquiring the required information from them. Alternatively, cyber-threat actors may find a way in through technical means, such as injecting a virus or using a spoofing network.

10.3.3 Sources of Cyber Threat

The main aim or motto of the cyber threat is to control the entire system. This controls can be taken over by any intruders, competitors, terrorist groups and others.

10.3.3.1 Terrorists

These sources are not mainly involved in destroying or controlling the system or organization by technical means. They look for traditional methods of attacking with arms and ammunition. Their goal is to rule over a place or system. Although, some terrorist groups who possess the technical skills and power will surely try to find loopholes with spying tools and techniques, as they need to learn the entire functioning of the system and work structure.

10.3.3.2 Insiders

Insiders includes people within an organization that have access to confidential data, can share it with other groups or competitors to obstruct the organization's growth. These insiders may also look for the destruction of such data, due to personal enmity. Nonetheless, above all, trust plays a major role in the growth of cyber threat.

10.3.3.3 Nations

Cyber threat today has no national boundaries. It is increasing with many confidential activities happening through the internet. Cyber-attacks between nations are a very dangerous activity which not only spoils cordial relations between the nations, but can also cut down communications, impact military strength and make life miserable for local citizens. Recently, many applications have been found collecting data from users' devices. These applications have been banned in most countries where they have discovered the risks of using such applications.

10.3.4 Cyber-Threat Environment

Cyber-threat actors require a space to conduct their malicious activities. This space, where all the malicious activities are performed by the cyber-threat actors, is known as the cyber-threat environment. The cyber-threat environment is mainly an online space. The devices that are used by the victims are mainly connected to the internet and this provides various opportunities for the cyber-threat actor to gain access to them and perform an attack. Cyber-threat actors mostly use a virtual private network that helps them to hide their identity and make them impossible to track.

10.3.5 Types of Cyber Threats

Advancements in technology have brought a lot of opportunities for cyber-threat actors to generate threats and attack victims. Some commonly performed cyber threats are mentioned below.

10.3.5.1 Botnets

A bot is a device that is infected with malicious software without the awareness of the user. This device is remotely controlled by a cyber-threat actor in order to impact the user activity or carry some other motto. Botnets are collection of such bots and have a coordination among them managed by the cyber-threat actor.

10.3.5.2 Denial of Service

This is a technique in which the cyber-threat actor disrupts the common activities of the user or victim by increasing the network traffic, or by making multiple requests to the server simultaneously with victim's requests. This can lead to a traffic flood. Due to this, the server is unable to determine the legitimate user request and cannot respond to it.

10.3.5.3 Man-in-the-Middle

This is also a technique through which the cyber-threat actor obstructs the communication between two entities (e.g., victim and a server). The victim is unaware of this actor being a part of the communication and therefore feels that the communication is directly with the secure server with no other party in the middle. As this cyber-threat actor is now a part of this communication, they can alter the communication, reroute the same or send malware along with the normal communication.

10.3.5.4 Password Cracking

This is an attempt to access an account with actual credentials. This is mainly done by insiders. Such cyber-threat actors are very well aware of the system and the victim's identity. As a lot of relevant information is already known to the cyber-threat actor, they can make use of the same and crack the password easily to gain control of confidential accounts or data.

10.3.5.5 Ransomware

This is a common cyber threat that deals with data withholding until some demands of the cyber-threat actors are fulfilled. Ransomware software restricts the targeted organization or the institution from accessing technical components such as servers, devices or workstations. One of the best delivery means of ransomware software is phishing, because of its easy-to-send ability. Phishing is an attempt at obtaining the sensitive information of a user by sending malicious content or by posing as a trusted sender of a communication. The type of mails that are shared and content present in the received mail should be closely monitored within an organization to identify malicious content mail.

10.4 Using Technologies to Mitigate Cyber Threats

Increasing cyber-threat sophistication and unexplored ways of attacking have caused alarm with concerns being raised regarding the capability of defending against cyber threats. Organizations and enterprises are common targets of venomous actors performing data breaches and creating operational losses. Such enterprises and organizations should understand how they can efficiently reduce cyber threats. A lot of technologies in today's era can help such organizations to tackle cyber threats and take precautionary measures. As the attackers are always one step ahead with new, unexplored ways of attacking, the existing solutions often fail to prevent attacks. Hence there is a need for anticipating possible novel attacks so that preventive solutions can be made available for zero-day attacks. The technologies, namely, machine learning, deep learning, artificial intelligence and blockchain can play an important role in mitigating existing cyber threats and also cater for zero-day attacks.

10.4.1 Comparison of Common Techniques

In this section, a lot of techniques are introduced to mitigate cyber-attacks and some common mitigation techniques are compared in Table 10.1. Out of the techniques the AI-based mitigation technique is most effective for cyber-attacks. Also, if two or more techniques are combined, they can provide better protection against cyber threats [11].

10.4.2 Artificial Intelligence

10.4.2.1 Introduction to AI in Cyber Security

Traditional technology put the security of the organization at risk. The reason for this is lack of data monitoring, and the inability to achieve data insights, and so on. The introduction of AI (AI) in cyber security will reduce the risks. Detecting anomalies in the system, analyzing network traffic, identifying possible threats in the system, and predicting the areas that may get attacked by cyber criminals, are all complex task that cannot be achieved by traditional security systems [12]. Artificial Intelligence is meant to be used for such complex and challenging tasks. With the use of AI and ML it is possible to prevent attacks before they occur. However, AI requires constant human interaction and training in the system as it uses ML and DL techniques. With the application of AI in cyber security, cyber threat can be reduced to a great extent.

10.4.2.2 Use of AI in Cyber Security

10.4.2.2.1 Exposing Cyber Threats

Cyber-threat actors follow a pattern or trend like other cyber criminals. An AI based system can keep a track of such patterns even though the pattern may be changed by the cyber-threat actors. This can help in recognizing the techniques that are likely to be used by cyber-threat actors to attack their victims.

TABLE 10.1

Comparison of Common Techniques to Mitigate Cyber Threats

Techniques	Description	Pros	Cons
Biometric	Based on physical and behavioral characteristics	– Efficient – Convenient – Identify fake entity from their physical or behavioral characteristics	– Impersonate characteristics – Limited to physical ability
Firewall	Helps to screen out viruses, malwares, etc.	– Access control – Prevent hackers – Block trojans	– Cannot block most of the malwares
Filtering tools	Tools that prevent users from visiting websites that have risk to online security.	– Can block emails – Can filter or block websites – Safeguards against ransomware	– Attackers can bypass the block system – Not efficient
Scanning software	Software that scans malicious files or virus files.	– Can alert the user about the malicious files. – Can scan efficiently	– If the alerts are ignored system may get attacked
ML-based	Machines are made to understand and analyze the pattern and learn from them to act accordingly on the changed behavior of the system.	– Efficient – Provide good results	– Complex in integration – Require some past knowledge for better learning of the machine
AI-based techniques	Adaptive learning technique	– Efficient – Provide better results – Adaptive	– Complex in integration
DL-based	Mimics the working of human brain and creates own features for recognition	– Efficient – Provide better results – Adaptive	– Very complex architectures – Computationally very costly
Blockchain	The distributed nature makes it impossible to be hacked.	– Distributed information – Immutable	– Very complex architectures – Computationally very costly

10.4.2.2.2 *Prediction of Breaching*

The AI technology can help in predicting the areas where breaching could happen. This will allow the organization to prepare themselves for allocation of appropriate tools and resources at these weak areas.

10.4.2.2.3 Response to Incidences

Furthermore, AI systems are capable of immediate responses to alerts. Thus, the system can prioritize responses in order to avoid future issues and reduce the system's vulnerabilities.

10.4.3 Machine Learning

Machine learning is a kind of model which is built by providing data to a machine, that is, a computer algorithm that can learn a pattern through the data and apply the learning in classifying or predicting unknown data. As ML is a powerful technique, it can be utilized by most organizations for their economic growth. On the other hand, ML can also be used by cyber criminals for their leverage. Cyber-threat actors can utilize these ML techniques that outsmart the defense system.

10.4.3.1 Use of Machine Learning in Reducing Cyber Threats

A lot of applications of ML can be made with proper techniques. With better infrastructure and resource availability, many ML applications have been launched by various organizations. Apart from ML being used for organizational data analysis and economic growth, ML can also be used in mitigating cyber threats. Countermeasures can be taken with the help of ML algorithms by analyzing attacks: domain of the attacks, intensity of the attacks and nature of the attacks. Some applications of ML are mentioned below that have been promising in reducing cyber threats.

10.4.3.1.1 Automated Security

Automated security can be achieved by using ML for identifying irregular traffic. Later, such suspicious activities can be programmed to be blocked automatically. Machine learning can also be used to identify anomalous behavior such as misused accounts or unauthorized access of the system. This can help in avoiding system vulnerabilities.

10.4.3.1.2 Advanced Antivirus Programs

Existing antivirus programs detect suspicious software by checking the signatures of the known malicious variants. This system fails in detecting malware that is not known or registered, and this can happen in a zero-day attack. This can be overcome with the help of ML that analyzes the source code or malware activity to identify authorized software.

10.4.3.1.3 Bane or Boon?

Machine learning, being a powerful technology, can be used by both cyber security experts and cyber criminals [13]. Machine learning can be used in training a machine to identify objects. The same technique is used in self-driving cars where machines are dependent on the algorithms for identifying obstacles. Also, as ML is capable of speeding up processes, it can help by thinking like human brain, where it can be used to guess passwords by learning from past datasets or login information. Machine learning is also capable of understanding the pattern of the human voice. The same technique can help cyber criminals in generating a model that mimic the victim's voice in cracking a voice-secured system, or sending abrupt voice messages on behalf of the victim.

Machine learning is a powerful tool that can be seen to assist in both offense and defense cyber activities. However, security experts should play an important role by being shoulder-to-shoulder with the advancements in antagonistic approaches of ML.

10.4.4 Deep Learning

Deep learning is a type of ML and could be called an advanced version of ML, where it uses deep networks for understanding the system.

The architecture of DL consists of some layers. The first layer is the input layer where the input is given. This input heads toward the second layer that consists of multiple functions. When an input passes through each of the layers, the fed input is changed to another input. All the changed input is now fed into an output layer where the entire network generates an output that is in the form of a prediction. There are a lot of DL frameworks that help in performing some DL experiments.

10.4.4.1 Deep Learning for Detection

10.4.4.1.1 Email Surveillance

Phishing activities are performed with mail. Nowadays, spam emails are already being detected, but the existing systems can be improved will the help of DL. Sometimes the email that is not meant to be a spam one gets classified as a spam mail in the existing system. This can be overcome with usage of the Natural Language Processing technique, where the emails can be analyzed and further classified as safe or unsafe.

10.4.4.1.2 Network Traffic Monitoring

Deep learning possesses the ability to handling multiple inputs and analyzing each and every input by a different method. This will ensure the classification of the network traffic is done with utmost accuracy and less false classification as compared to typical ML [14]. Once the traffic is analyzed and classified, it can be used to detect suspicious traffic and block the same to protect the system from any cyber-attack. This can also be used in predicting the area where the attack seems likely to happen.

10.4.5 Blockchain

10.4.5.1 What Is Blockchain?

Blockchain is a decentralized distributed ledger system that provides security of data. Blockchain provides a truly secure system where the transactions can be verified and validated without the involvement of the third party. Basically, blockchain is a combination of various technologies. It is developed on a platform by using protocols. Also, it is on a peer-to-peer network system containing records and most importantly uses private key cryptography for identification. The work of securing the system is intrinsic, provided by the simple, robust and yet sophisticated architecture of blockchain. The blockchain's capability to provide data confidentiality, integrity and availability can help to mitigate cyber threats.

10.4.5.2 Nature of Blockchain

Blockchain allows transactions to occur in a secured manner, without the intervention of third party in the same. This is achieved with the use of a digital signature for verification of the parties involved in the transaction. With the use of such digital signatures there are far fewer chances for cyber-threat actors to tamper with the data without the digital signature. Moreover, blockchain uses the hash encryption of data to secure the information. After verification, the encrypted information is added to the blockchain. This hash encryption is nearly impossible to crack, making it a suitable method in terms of cyber security.

In blockchain, for every data change a new hash is generated and add to the chain; this ensures that the data is recent and updated. Also, if any tampering to the data is done, it can be easily detected, as the hash value change will reflect it immediately.

10.4.5.3 Use of Blockchain in Data Integrity

Blockchain is known for its capability for data integrity. Blockchain is immutable, which means that once the transaction has been confirmed, it cannot be altered by any means, thus, ensuring the integrity of the transaction data is maintained throughout. Blockchain protects the system from attackers, reduces fraud and lowers the chances of data alterations or data stealing [15]. All this can be achieved due to blockchain's distributed nature. Every system that is processing the blockchain needs to be attacked in order to corrupt the entire system. Also, the number of systems in a blockchain may go beyond a thousand, and therefore, this makes it impractical for an attacker to attack the data.

10.4.5.4 Use Cases of Blockchain in Cyber Security

10.4.5.4.1 Decentralized Storage

Data has always been an important factor of organizational growth and strength. If any attack is made to the data, both the growth and strength of the organization may be impacted. With the use of a decentralized block-chain solution, organizations can store data as well as make it more secure. This is achieved as the decentralized blockchain platform breaks the data and distributes it across multiple network nodes, which makes it impractical to steal it or tamper with it.

10.4.5.4.2 Securing DNS

A cyber-threat actor's target is mainly a Domain Name System (DNS). These attackers try to exploit the connection between the site and the IP address. The main aim behind doing this is to crash the site. Site crashing can cause temporary shutdown of service, due to which the organization or the company may suffer a huge loss. Blockchain can provide a powerful solution to such attacks. With the decentralized nature of blockchain, the DNS information can be stored in a distributed system. Along with this, all the connections can be made immutable with the help of smart contracts.

By deeper study of these technologies, it is recognized that we can create a cyberinfrastructure using all these technologies to build an automated system capable of mitigating cyber threats and attacks.

10.5 Cyberinfrastructure

Cyberinfrastructure is an environment that supports advanced technology for data acquisition, data management, data visualization, data mining and various computation and processing services distributed over the internet, beyond single organizational or institutional scope [16].

The proposed solution has four parts. The first part is to aggregate the data from various nodes in the network regarding cyber-criminal activities using AI techniques namely information agents. The second part is to mine the aggregated data further to extract meaningful data highlighting possible criminal activities using ML techniques. The third part is to build knowledge for classification using DL techniques that identify the threats and the system vulnerabilities intelligently and futuristically so that cyber threats and attacks can be mitigated. The last part is to secure all collected insights so that the attackers cannot tamper with them.

Figure 10.1 contains a flow diagram of how the integration of artificial intelligence, machine learning deep learning and blockchain can be made at various areas of a cyberinfrastructure to mitigate cyber threats.

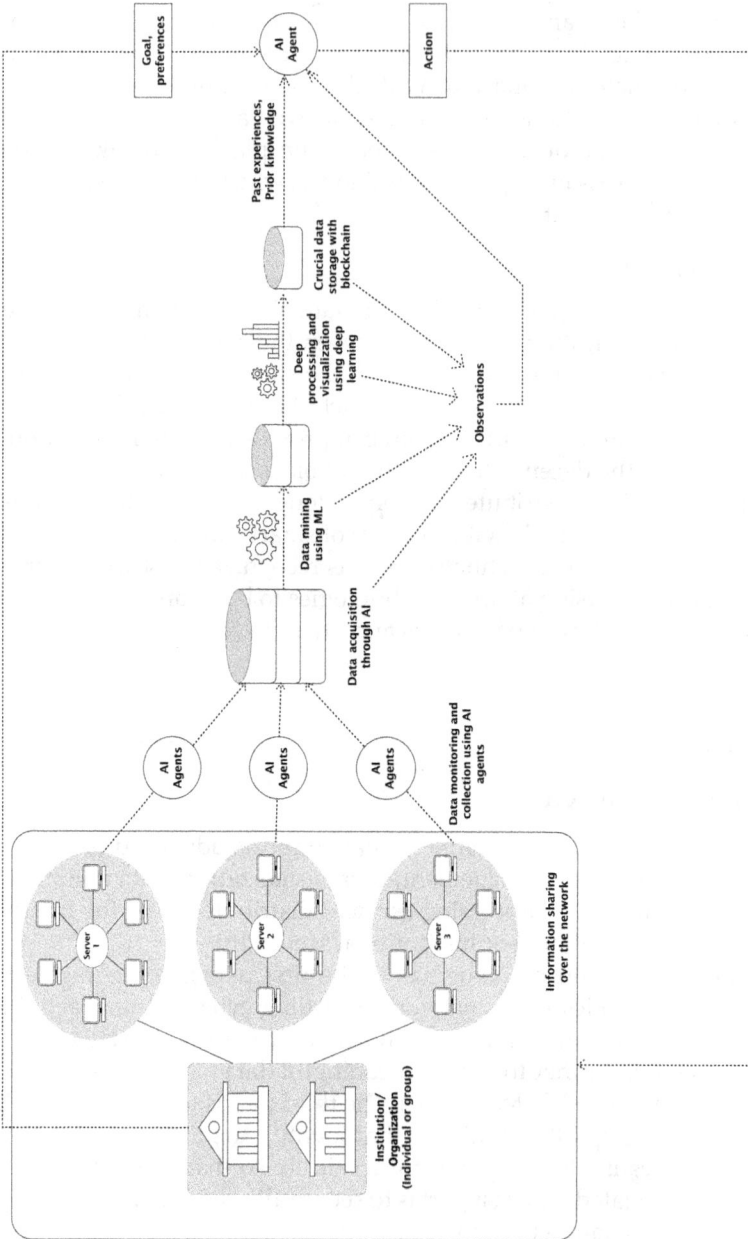

FIGURE 10.1
Cyberinfrastructure for cyber-threat mitigation.

Groups of institutes or organizations that share data or communicate through a network should be monitored carefully. This can be achieved with the help of AI agents that can perceive and observe these data flows over the network. Any anomalous behavior over the shared network will be detected by the AI agents and necessary actions will be taken. If the anomalous behavior is not detected with certainty, then the data acquired by these AI agents will be mined using ML algorithms for extracting the abnormal activity details from the normal running environment. This will further be processed and analyzed by DL technology to gain important insights from the acquired data.

Data insights and observations are crucial, when it comes to the actions that are dependent on these insights. If cyber-threat actors target the collected insights, then the AI agent will receive the false data and may fail to recognize the attack and may not take action over the attacks in real time. Such crucial data should be transacted over the network with the use of blockchain technology. Smart contracts can be used to ensure that the data received and stored from the network is authenticated and integrity is maintained. This will ensure that the processed data received by the AI agents is not tampered with.

All these observations at different stages will be collectively sent to the AI agent that already contains the goal and preferences of the organization/institution, to make a list of actions that need to be taken on the perception of observations from various stages. With all the required data, the AI agent will be able to take necessary action such as blocking the threat files, removing the unknown entity, restricting the area from processing further, and so on, whenever a cyber-threat is detected in the environment. The AI agents can also adapt themselves to these new attacks so that they can self-sustain action on these attacks in the future.

10.6 Conclusion

To mitigate cyber threats, the cyberinfrastructure of any organization/institution should be strong and well designed with advanced technologies that can self-learn and prepare themselves for identification and/or prevention of unknown attacks. In this work, we defined a design of cyberinfrastructure with advanced technologies such as AI, ML, DL and blockchain that can be used to deal with cyber-attacks. These technologies, when used together, can be used to build a robust, self-sustained cyberinfrastructure.

10.7 Future Scope

The proposed solution depends hugely on the real-time environment data that undergoes such cyber-attacks for its performance testing and, hence, needs to be tested in the real environment. As future scope, we plan to study ways of optimizing the costs and computational power at various stages to improve and balance the entire cyberinfrastructure. Furthermore, we are interested in defining models that can be used to secure individual devices from cyber threats.

References

[1] F. O. Olowononi, D. B. Rawat, and C. Liu, "Resilient Machine Learning for Networked Cyber Physical Systems: A Survey for Machine Learning Security to Securing Machine Learning for CPS," *IEEE Communications Surveys & Tutorials*, !–1, 2020. doi:10.1109/comst.2020.3036778

[2] K. Sathya, J. Premalatha, and S. Suwathika, *International Conference on Communication and Signal Processing*, July 28–30, 2020, India 978-1-7281-4988-2/20/$31.00 ©2020 IEEE.

[3] G. Apruzzese, M. Colajanni, L. Ferretti, A. Guido, and M. Marchetti, "On the Effectiveness of Machine and Deep Learning for Cyber Security," *2018 10th International Conference on Cyber Conflict (CyCon)*, 2018, pp. 371–390, doi:10.23919/CYCON.2018.8405026

[4] S. Bagui, D. Nandi, S. Bagui, and R. J. White, "Classifying Phishing Email Using Machine Learning and Deep Learning," *2019 International Conference on Cyber Security and Protection of Digital Services (Cyber Security)*, 2019. doi:10.1109/cybersecpods.2019.8885143

[5] M. Roopak, G. Yun Tian, and J. Chambers, "Deep Learning Models for Cyber Security in IoT Networks," *2019 IEEE 9th Annual Computing and Communication Workshop and Conference (CCWC)*, 2019, pp. 0452–0457. doi:10.1109/CCWC.2019.8666588

[6] G. Bendiab, S. Shiaeles, A. Alruban, and N. Kolokotronis, "IoT Malware Network Traffic Classification Using Visual Representation and Deep Learning," *2020 6th IEEE Conference on Network Softwarization (NetSoft)*, 2020. doi:10.1109/netsoft48620.2020.9165381

[7] A. Farion, O. Dluhopolskyi, S. Banakh, N. Moskaliuk, M. Farion, and Y. Ivashuk, "Using Blockchain Technology for Boost Cyber Security," *2019 9th International Conference on Advanced Computer Information Technologies (ACIT)*, 2019. doi:10.1109/acitt.2019.8780019

[8] H. M. Farooq and N. M. Otaibi, "Optimal Machine Learning Algorithms for Cyber Threat Detection," *2018 UKSim-AMSS 20th International Conference on Computer Modelling and Simulation (UKSim)*, 2018. doi:10.1109/uksim.2018.00018

[9] I. Sharafaldin, A. H. Lashkari, and A. A. Ghorbani, "Toward Generating a New Intrusion Detection Dataset and Intrusion Traffic Characterization," *ICISSP*, 2018, pp. 108–116.

[10] R. Vijayanand, D. Devaraj, and B. Kannapiran, "Intrusion Detection System for Wireless Mesh Network Using Multiple Support Vector Machine Classifiers with Genetic-Algorithm-Based Feature Selection," *Computers & Security*, vol. 77, pp. 304–314, 2018.

[11] Fatima Salahdine and Naima Kaabouch, "Social Engineering Attacks: A Survey," *Future Internet*, vol. 11, 2019. doi:10.3390/fi11040089

[12] B. S. Sagar, S. Niranjan, N. Kashyap, and D. N. Sachin, "Providing Cyber Security Using Artificial Intelligence – A Survey," *2019 3rd International Conference on Computing Methodologies and Communication (ICCMC)*, 2019, pp. 717–720. doi:10.1109/ICCMC.2019.8819719

[13] The Seventh International Conference on Data Analytics 2018, Manjeet Rege, Graduate Programs in Software University of St. Thomas, St. Paul, MN, USA, Email: rege@stthomas.edu. Raymond Blanch K. Mbah, Graduate Programs in Software University of St. Thomas, St. Paul, MN, USA, Email: kong1343@stthomas.edu. ISBN: 978-1-61208-681-1

[14] G. Nguyen, S. Dlugolinsky, V. Tran, and Á. López García, "Deep Learning for Proactive Network Monitoring and Security Protection," *IEEE Access*, vol. 8, pp. 19696–19716, 2020. doi:10.1109/ACCESS.2020.2968718

[15] S. Patil, S. Kadam, and J. Katti, "Security Enhancement of Forensic Evidences Using Blockchain," *Third International Conference on Intelligent Communication Technologies and Virtual Mobile Networks (ICICV)*, 2021, pp. 263–268. doi:10.1109/ICICV50876.2021.9388486

[16] C. A. Stewart, S. Simms, B. Plale, M. Link, D. Y. Hancock, and G. C. Fox, "What Is Cyberinfrastructure," *Proceedings of the 38th Annual Fall Conference on SIGUCCS – SIGUCCS '10*, 2010. doi:10.1145/1878335.1878347

11

Quantum-Safe Cryptography

Kunal Mohan Meher

Associate Professor, Xavier Institute of Engineering, Mumbai, India

CONTENTS

11.1 Introduction

A traditional PC processes on bits that are in one of two states: "1" or "0" (additionally called "high" or "low", "on" or "off", "true" or "false"). A quantum PC processes on qubits with possible states "1" or "0", or in limitlessly numerous superposition conditions of "1" and "0". In a quantum PC, after calculation, the qubits are estimated. That powers qubits which are in superposition to snap into one of the two states "1" or "0" by a specific likelihood, relying upon their superposition state.

The infinitely large space of superposition states during computation, and the entanglement of qubits, allow quantum computers to solve certain classes of problems much faster than traditional computers. In particular, hard problems faced in today's cryptosystems suddenly become feasible when a quantum computer is used. Companies like Google, IBM, Intel and Rigetti are working to build a powerful quantum computer. A 72-qubit quantum machine has already been built. There are two key schemes of quantum processing that affect the strength of cryptographic algorithms: Grover's algorithm and Shor's algorithm. Although Grover's algorithm running on a powerful quantum computer is a threat to symmetric key algorithms, the easy solution is to increase key size. SHA-256 and AES-128 are both quantum-safe as indicated by the assessment measures in the National Institute of Standards and Technology (NIST) post-quantum cryptography standardization project.

While quantum cryptography portrays utilizing quantum strengths at the center of a security system, post-quantum cryptography (sometimes referred to as quantum-resistance, quantum-proof or quantum-safe) alludes to cryptographic calculations (predominantly public-key calculations) that are believed to be secure against an assault by a quantum PC [1]. The most popular illustration of quantum cryptography is quantum key distribution (QKD). Post-quantum cryptography (PQC) is tied in with getting ready for the time of quantum processing by refreshing existing numerical-based guidelines and calculations. Cryptographic researchers are exploring post-quantum cryptography to give choices utilizing strong numerical issues which can't be broken by quantum PCs. All things considered, both QKD and PQC calculations will discover their applications later on in a post-quantum cryptographic world.

11.2 Current State of Cryptosystems

In the digital world, cryptography is commonly associated with three main principles: confidentiality, integrity and authentication. These principles provide an assurance that information is trustworthy and can only be accessed by authorized users. Each principle is underpinned by the implementation of cryptographic functions. There are three broad categories of cryptosystems (i.e., cryptographic algorithms) hash functions, symmetric-key algorithms and asymmetric-key algorithms. Hash functions are used in digital signature, password protection, pseudo random generation (PRG) and randomness extraction such as hash-based key derivation function (HKDF). Symmetric-key algorithms are used for encryption, message authentication code (MAC) and generating deterministic random numbers. Asymmetric-key cryptosystems are used for public key encryption, computing digital signature and establishing cryptographic keying material.

The asymmetric algorithms in use today are vulnerable to a powerful quantum computer. The hash algorithms and symmetric key algorithms with larger key size are considered to be quantum-safe. Today, various protocols like Secure Socket Layer (SSL)/Transport Layer Security (TLS), Signal, Internet Key Exchange (IKE), (Secure Shell) SSH and Secure/Multipurpose Internet Mail Extensions (S/MIME) are used in the vulnerable internet for secure communications. The state-of-the-art cryptographic algorithms used for confidentiality (also providing authentication) in the protocols include Authenticated Encryption (AE) Schemes such as Advanced Encryption

Standard using Galois Counter Mode (AES-GCM), AES using Counter with CBC-MAC (AES-CCM) and ChaCha20-Poly1305. ChaCha20-Poly1305 is an Authenticated Encryption with Additional Data (AEAD) cipher. The algorithms used for integrity include cryptographic hash functions such as Secure Hash Algorithm (SHA-256). The digital signature algorithms such as Elliptic Curve Digital Signature Algorithm (ECDSA)/Edwards-curve Digital Signature Algorithm (EdDSA) have become more popular. The key exchange algorithms such as Elliptic Curve Diffie-Hellman Empheral (ECDHE) using Curve 25519 or 448, Extended Triple Diffie-Hellman (X3DH) and the key management algorithms such as Double Ratchet algorithm are used for securely generating secret keys at both ends. It can be noted that RSA key exchange, DES cryptosystem, SHA-1 and MD5 hash functions are all depreciated. Of the above-mentioned algorithms, ECDSA, EdDSA, ECDHE, X3DH and Double Ratchet algorithm are not quantum-proof, while AES, ChaCha20 and SHA-256 are considered to be quantum-proof [2].

11.2.1 Security Issues with Current Cryptosystems

There is a notable risk involved if we delay transition to quantum-safe cryptography for the reasons explained below:

- Because powerful quantum computers will be a reality in future, secret information today, in store or transit, is insecure. Attackers may somehow capture encrypted confidential data now. They are not able to decrypt it today by using traditional computers. But once they have access to powerful quantum computers in future, attackers will be able to decrypt it and make sense of information out of it. This means that today's secret encrypted data may not be confidential once quantum computers become a reality.

- There is a danger that digitally signed data today may not be reliable in the future. If users use a structure that digitally signs data today with a non-quantum resistant digital signature, an adversary with a quantum computer in the future could alter the signature or repudiate ever signing certain information. The chain of faith would be lost. When quantum PCs can break signature schemes, the dangers are broad. For instance, a programmer could break a Windows programming update key and send counterfeit updates (malware) to a PC. Hence, the need to implement quantum-resistant cryptography is not relegated to sometime in the future, but is of real import today.

- Mainstream protocols utilized on the internet for secure correspondence, like SSH, SSL/TLS, IKE, S/MIME and Signal, use public key cryptography for key exchange, authentication and digital signature. With advances in quantum computing, all these protocols will be vulnerable.

11.3 Current State of Post-Quantum Cryptography (PQC)

There are two operations that use public key cryptography: key establishment and digital signatures. Within key establishment, there are two common methods: key agreement and key transport.

There are five leading groups of PQ primitives – code-based, lattice-based, isogeny-based, hash-based and multivariate-based.

The National Institute of Standards and Technology (NIST) is working on standardization of PQC algorithms. In July 2020, NIST shortlisted seven algorithms as third round finalists, as listed in Table 11.1 In addition, eight more candidate algorithms are under consideration for Round 3 as listed in Table 11.2 [3].

TABLE 11.1

NIST Third Round Finalists

Scheme	Enc/Sig	Family	Hard Problem
Classic McEliece	Enc	Code-Based	Decoding random binary Goppa codes
Crystals-Kyber	Enc	Lattice-Based	Cyclotomic Module – LWE
NTRU	Enc	Lattice-Based	Cyclotomic NTRU Problem
Saber	Enc	Lattice-Based	Cyclotomic Module – LWR
Crystals-Dilithium	Sig	Lattice-Based	Cyclotomic Module – LWE and Module – SIS
Falcon	Sig	Lattice-Based	Cyclotomic Ring – SIS
Rainbow	Sig	Multivariate-Based	Oil-and-Vinegar Trapdoor

TABLE 11.2

NIST Third Round Alternate Candidates

Scheme	Enc/Sig	Family	Hard Problem
BIKE	Enc	Code-Based	Decoding quasi-cyclic codes
HQC	Enc	Code-Based	Coding variant of ring – LWE
Frodo – KEM	Enc	Lattice-Based	LWE
NTRU – Prime	Enc	Lattice-Based	Non-cyclotomic NTRU Problem or Ring – LWE
SIKE	Enc	Isogeny-Based	Isogeny problem with extra points
GeMSS	Sig	Multivariate-Based	'Big-Field' Trapdoor
Picnic	Sig	Symmetric Crypto	Preimage resistance of a block cipher
SPHINCS+	Sig	Hash-Based	Preimage resistance of a hash function

11.4 Challenges in Post-Quantum Cryptography (PQC)

It is observed that migration from one cryptographic primitive used in the protocol over the internet to another primitive is a very slow process. It has required more than 20 years to establish our cutting-edge public key cryptography framework. It will take critical endeavors to guarantee a protected and smooth movement from the current, broadly utilized, cryptosystems to their quantum-safe counterparts. Thus, independent of whether we can predict the exact time of the creation of a powerful quantum computer, we should immediately start to set up our data security frameworks for quantum-resistance. A huge worldwide community has arisen focused on discovering the issues of data security in a quantum-computing future. The people group is attempting to distinguish and address the constituent difficulties in PQC movement with the expectation that our public key framework may stay unblemished by using new quantum-safe algorithms.

The following are the limitations in PQC deployment (to replace today's modern primitives):

- The security of existing post-quantum schemes cannot be fully verified (as a powerful quantum computer has not yet been built to test it).

- We sometimes fail to consider possible attacks unique to a quantum adversary. In other words, the classical definition of security may not capture the right notion of security in the presence of quantum attackers.

- Quantum-safe suppositions made today won't clearly suggest the quantum security of an algorithm, because of other basic issues that could be unobtrusive and barely noticeable. Security evidence may totally fall through within the sight of quantum assaults.

- When comparing post-quantum cryptography and the present public key cryptosystems, it can be discovered that post-quantum cryptography, generally, has bigger key and signature sizes and needs more computing and memory. This results in remarkably larger amounts of data that need to be sent over a communications link for key establishment and signatures. These larger key sizes also require more storage inside a device. However, while key sizes are larger, most quantum-resistant algorithms are more computationally efficient than existing public key algorithms. The PQC natives are exceptionally useful for everything with the exception of maybe extremely obliged Internet of Things gadgets and radio.

11.5 Approaches for Post-Quantum Cryptography (PQC) Migration

It is not feasible to replace existing algorithms in the protocol with their post-quantum counterparts, considering the security of PQ algorithms cannot be verified fully on traditional computers today. Given that the NIST will take some years for PQC standardization and above limitations, fundamentally, two feasible approaches exist to deal with the issue: a hybrid scheme and proactive measures for pre-quantum cryptography.

11.5.1 Hybrid Scheme

A hybrid algorithm is a mixing of a traditional and a post-quantum algorithm, meaning that the final algorithm is, at the minimum, as safe as one of the algorithms used. The use of hybrid algorithms can secure several kinds of future threats and dangers. It is strongly proposed to ease migration to the post-quantum epoch. In addition to a primary goal of security, hybrid schemes could also achieve additional objectives such as backward compatibility, high performance, low latency, no extra round trips and no duplicate information. But, there is a need to design hybrid solutions considering issues like negotiation, selecting the number of component algorithms, ways of conveying cryptographic data and how to combine cryptographic data [4].

Hybrid key exchange (Hybrid KEM): This means that the session key should remain secure (and thus application data confidential) as long as one of the ingredient key exchange mechanisms is unbroken. There are different approaches for hybrid key encapsulation such as HKDF_then_XOR, XOR_then_HKDF and Concat_then_HKDF which are used to combine keys generated by traditional and post-quantum algorithms [5, 6]. For instance, Google experimented with using a hybrid of an Elliptic Curve key agreement along with a Ring Learning with Errors key agreement into the Google Chrome Canary browser. There is another approach of passing one of the secret keys as salt value to HKDF, and another secret key as input to HKDF for deriving session key(s) [7].

Figure 11.1 shows a hybrid key encapsulation mechanism by applying hash-based KDF on master key from a post-quantum KEM algorithm, and passing master key from a traditional KEX algorithm as salt value. Similarly, Figure 11.2 shows a hybrid key encapsulation mechanism by applying hash-based KDF on master key from a traditional KEX algorithm and passing master key from a PQ KEM algorithm as salt value. These methods are effective as they require only one KDF computation and this avoids the need for XOR operation or concatenation operation.

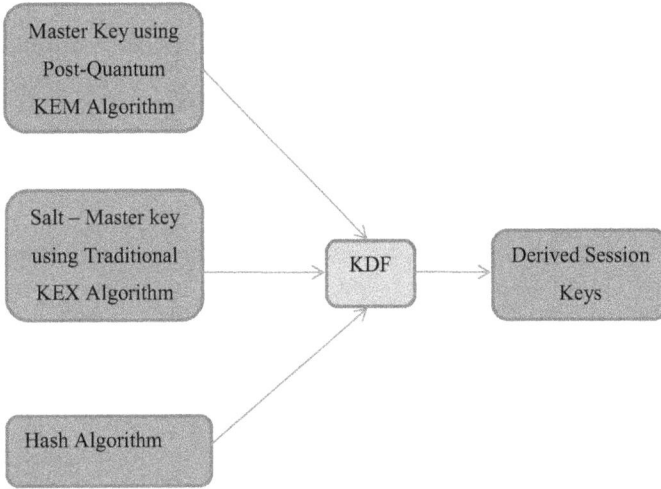

FIGURE 11.1
Traditional secret key as salt to HKDF.

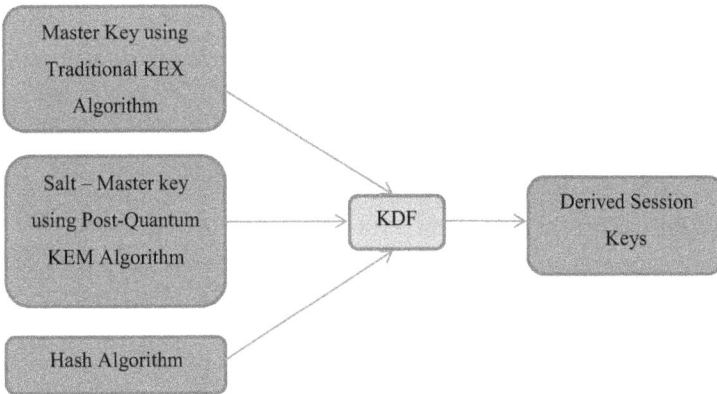

FIGURE 11.2
PQ secret key as salt to HKDF.

Hybrid digital signature: This means that the protocol should provide secure authentication as long as one of the digital signatures schemes is unbroken at the time of session establishment. There are different approaches to combine two digital signatures: traditional signature and post-quantum signature such as concatenation, weak-nesting and strong-nesting. A lot of research is going on to find best way to combine digital signatures.

- Concatenation: s1 = Sign1(M) and s2 = Sign2(M).
- Weak-Nesting: S1 <- Sign1(m) and S2 <- Sign2(S1)
- Strong-Nesting: S1 <- Sign1(m) and S2 <- Sign2((m, S1))

In the above notations, Sign1 and Sign2 can be a traditional digital signature or a PQ digital signature, but Sign1 and Sign2 should not be of the same type at any given instance [8–10].

Nonetheless, post-quantum primitives face the problem of bigger signature and key sizes. So, until the post-quantum primitives are standardized, the following approaches can be considered to deal with bigger signatures and key size:

1. Signatures with Message Recovery

   There is an approach which uses traditional signature with full message recovery to deal with bigger signature and key size for PQ transition as mentioned below:

   $$S1 \leftarrow PQ\_Sign(m) \text{ and } S2 \leftarrow Traditional\_Sign\_with\_message\_recovery(S1)$$

   The channel bandwidth is saved as only a small-sized traditional signature needs to be sent to the receiver, as shown in Figure 11.3. The bigger PQ signature need not be sent and recovered at the receiver side [11]. Figure 11.4 shows the signature verification in case of combined signature with message recovery.

2. Adjustments of the Protocol to Use KEMs for Authentication

   In this approach, Key Encapsulation Mechanism is used instead of digital signature for authentication which is implemented in KEM-TLS and PQ Wireguard

3. Stateful Hash-Based Signatures

   Stateless hash-based signatures like Sphincs, which are under NIST standardization, have bigger signature sizes. So, stateful hash-based signatures like LMS and XMSS can be considered [12, 13].

Hybrid encryption: This means that the encrypted message remains secure as long as one of the encryption algorithms remains secure. There are two approaches possible for hybrid encryption to encrypt plaintext (PT) to get ciphertext (CT):

- CT = Traditional Encryption (PQ Encryption (PT))
- CT = PQ_Encryption (Traditional Encryption (PT))

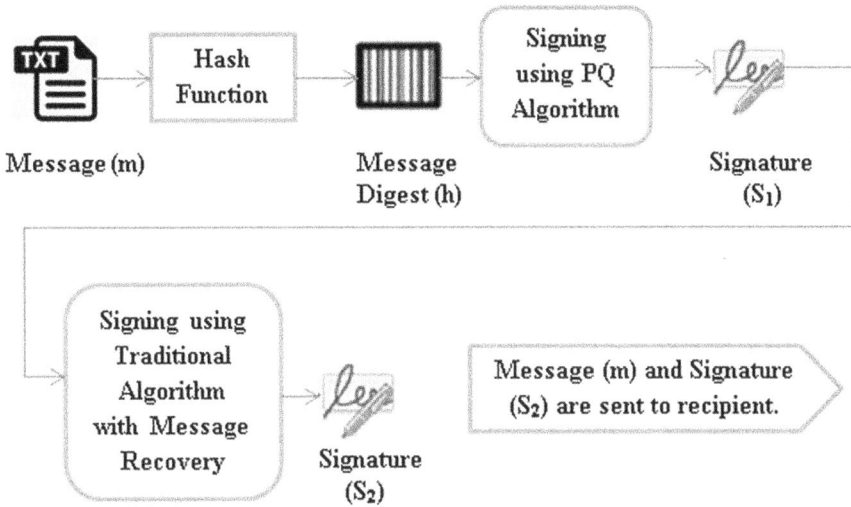

FIGURE 11.3
Message signing in combined signature using message recovery.

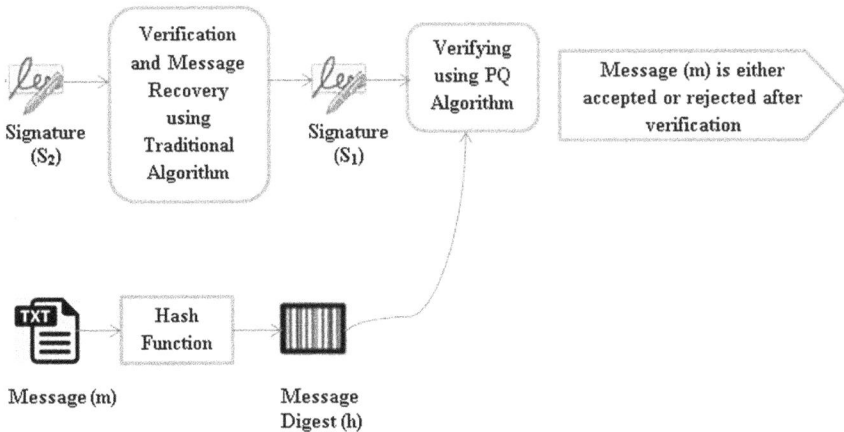

FIGURE 11.4
Signature verification in combined signature using message recovery

11.5.2 Protective Measures for Pre-Quantum Cryptography

Another choice is to employ the theoretically simple, but organization-ally complex, computation of mixing pre-shared keys into all keys con-firmed via public key cryptography. The users who do not want to deploy PQC primitives before standardization can safeguard their setup by mentioning preserved shared secret data in the key derivation, as well as the key matter obtained by a public key functionality [14, 15].

References

[1] Kunal Meher and Divya Midhunchakkaravarthy. 2020. Hybrid Solution (ECDHE + NewHope) for PQ Transition. *International Journal of Engineering and Advanced Technology*, Volume 9, Issue 2, pp. 3893–3894. https://doi.org/10.35940/ijeat. B3799.129219

[2] Kunal Meher and Divya Midhunchakkaravarthy. 2020. The State-of-the-Art Cryptographic Algorithms", *Journal of University of Shanghai for Science and Technology*, Volume 22, Issue 10, pp. 142–145. https://jusst.org/wp-content/uploads/2020/10/The-State-of-the-art-Cryptographic-Algorithms.pdf

[3] NIST. 2020. PQC Standardization Process: Third Round Candidate Announcement. https://csrc.nist.gov/News/2020/pqc-third-round-candidate-announcement. (Accessed August 16, 2021).

[4] Eric Crockett, Christian Paquin, and Douglas Stebila. 2019. Prototyping Post-Quantum and Hybrid Key Exchange and Authentication in TLS and SSH, pp. 1–24. https://eprint.iacr.org/2019/858.pdf

[5] F. Giacon, F. Heuer, and B. Poettering. 2018. KEM Combiners, pp. 1–29. https://eprint.iacr.org/2018/024.pdf

[6] N. Bindel, J. Brendel, M. Fischlin, B. Goncalves, and D. Stebila. 2019. Hybrid Key Encapsulation Mechanisms and Authenticated Key Exchange. *10th International Workshop on Post-Quantum Cryptography*. https://eprint.iacr.org/2018/903.pdf

[7] Kunal Meher and Divya MidhunChakkaravarthy. 2021. New Approach to Combine Secret Keys for Post-Quantum (PQ) Transition. *Indian Journal of Computer Science and Engineering*, Volume 12, Issue 3, pp. 629–633. https://doi. org/10.21817/indjcse/2021/v12i3/211203138

[8] Nina Bindel, Udyani Herath, Matthew McKague, and Douglas Stebila. 2017. Transitioning to a Quantum-Resistant Public Key Infrastructure. https://eprint. iacr.org/2017/460.pdf

[9] Kunal Meher and Divya Midhunchakkaravarthy. 2021. Best-Fit Dual Signatures for PQ Transition. *4th Biennial International Conference on Nascent Technologies in Engineering (ICNTE), 2021*, pp. 1–3 https://doi.org/10.1109/ICNTE51185.2021.9487705

[10] Kunal Meher and Divya MidhunChakkaravarthy. 2021. Approaches to Deal with Bigger Signatures for Post-Quantum Transition. Presented in *International Conference on Emerging Trends in Engineering and Technology*.

[11] Ward Beullens, Jan-Pieter D'Anvers, Cyprien de Saint Guilhem et al. 2021. Post-Quantum Cryptography – Current State and Quantum Mitigation, ENISA Report.

[12] David McGrew, Panos Kampanakis, Scott Fluhrer, Stefan-Lukas Gazdag, Denis Butin, and Johannes Buchmann. 2017. State Management for Hash-Based Signatures. https://eprint.iacr.org/2016/357.pdf

[13] Furqan Shahid, Iftikhar Ahmad, Muhammad Imran, and Muhammad Shoaib. 2020. Novel One Time Signatures (NOTS): A Compact Post-Quantum Digital Signature Scheme. *IEEE Access*, Volume 8, pp. 15895–15906. https://doi.org/10.1109/ACCESS.2020.2966259

[14] D. Sikeridis, P. Kampanakis, and M. Devetsikiotis. 2020 *Post-Quantum Authentication in TLS 1.3: A Performance Study. Network and Distributed Systems Security (NDSS) Symposium*, USA. https://eprint.iacr.org/2020/071.pdf

[15] Peter Schwabe, Douglas Stebila, and Thom Wiggers. 2020. Post-Quantum TLS Without Handshake Signatures. Session 5B: Secure Messaging and Key Exchange, USA, pp. 1461–1480. https://dl.acm.org/doi/pdf/10.1145/3372297.3423350

Index

Pages in *italics* refer to figures and pages in **bold** refer to tables.

For Product Safety Concerns and Information please contact our EU
representative GPSR@taylorandfrancis.com
Taylor & Francis Verlag GmbH, Kaufingerstraße 24, 80331 München, Germany